THE LAST AMERICAN

THE REMARKABLE LIFE OF
JOHN GLENN

ALICE L. GEORGE

CHICAGO
REVIEW
PRESS

Copyright © 2021 by Alice L. George

All rights reserved

First edition

Published by Chicago Review Press Incorporated

814 North Franklin Street

Chicago, Illinois 60610

ISBN 978-1-64160-213-6

Library of Congress Control Number: 2020942075

Print book interior design: Jonathan Hahn

Printed in the United States of America

5 4 3 2 1

To the men and women who risk their lives in opening the way to exploration of the final frontier.

CONTENTS

PROLOGUE

Without heroes, we are all plain people,
and don't know how far we can go.
—Bernard Malamud

Hero.

John Glenn absorbed the title like a plant consuming a drop of water. It became a part of him. It nourished him, stimulated him, invigorated him. It changed his life without displacing his essence. "John Glenn, the hero" would thrive, but "John Glenn, the small-town boy" would always rule his heart. In an age almost devoid of heroes, he became an anomaly, the last of his kind in an antiheroic age—a man with both physical courage and moral conviction. In old age, he witnessed the decline of honesty, courage, and empathy in American discourse. He saw the abandonment of heroes, who were replaced by visions of villains on all sides in a polarized America. However, he never relinquished the exalted status bestowed upon him by the American people, who had followed his life and found something great within it. His heroic stature was rare at the start of the twenty-first century, but others of his breed hold places in the nation's history.

This biography does not strive to place Glenn on a pedestal; instead, it sets out to show how a human being, flawed like all of us, was able to navigate a remarkable life. Fueled by hard work, bravery, love, and devotion, he assembled a long list of accomplishments. He had no superpowers and wore no halo, and still, his lifetime of achievements turned him into a venerated figure—a role he never sought. Throughout his life, he repeatedly greeted challenges by stepping forward and willingly offering his life in service to his country. Embracing his humanity, he aimed to lead a good life. At that he succeeded, providing an inspirational model for other Americans to follow in trying times.

His life was not only exemplary but also truly phenomenal. He married the only girl he had ever dated, and his love never wavered during more than seventy years together. He fought in two wars. During the first, he met his childhood hero, Charles Lindbergh; during the second, he flew alongside one of the nation's greatest sports stars, Ted Williams, and downed three Soviet MiGs in the war's final nine days. In the late 1950s, he set an aviation speed record and won the top prize on a TV game show. Then he soared into space as the first American to orbit Earth. He became a close friend to presidential candidate Robert F. Kennedy and played a key role in the aftermath of his assassination. Always pursuing new challenges, Glenn served as a US senator for more than two decades before becoming the oldest person to travel in space.

In a life full of big days, February 20, 1962, the day of Glenn's first orbital flight during Project Mercury, affected his public life most profoundly. Without surrendering to fear, he flew into a maelstrom of unknowns. Afterward, all fifty states from Maine to Hawaii celebrated, with ticker tape raining in New York City and teary-eyed lawmakers saluting him on Capitol Hill. The Soviet Union had beaten the United States to every major space spectacular—first satellite (*Sputnik* in 1957), first manned flight (Yuri Gagarin in April 1961), first manned orbit (also Gagarin)—but Glenn's flight, which, unlike secret Soviet flights, had garnered a full day of television coverage, had elevated American spirits, while making each citizen part of a heart-pounding space opera. "He became the American space hero. . . .

He was the guy who came to define that image of the astronaut," said Michael Neufeld, a curator at the National Air and Space Museum.[1]

From the first days after his flight, Glenn relaxed in the warmth of the spotlight. Confident without being cocky, he seemed comfortable with fame, and at the same time, both humble and generous. His short addresses as a hero were far more powerful than the often-wooden, overly detailed speeches he delivered later in life as a senator and a presidential candidate. Glenn never showed any sense of being oppressed by crowds of fans, and for more than half of his life, he unfailingly showed great willingness to stop and sign autographs. His apparently easy transition into life as a celebrity surprised many observers, but its naturalness added to Americans' respect for him.

In *The Right Stuff*, Tom Wolfe wrote that President John F. Kennedy's fans screamed and grasped souvenirs, while the masses "anointed" Glenn and the Mercury astronauts with their tears.[2] Demonstrating both his valor and his virtue, Glenn made no secret of his desire to embody what was best about America. The *New York Times* called him "fabulously courageous,"[3] and *Time* reported, "Glenn's modesty, his cool performances, his dignity, his witticisms, his simplicity—all caught the national imagination."[4]

Over the course of a long life, Glenn gave to his country again and again. Through twenty-three years in the military and twenty-four years in the US Senate, he literally devoted most of his adult life to his country and was still on the job at age seventy-seven. More than a daredevil pilot, this man was brave enough to ride a rocket—twice—and self-assured enough to listen to and truly hear Americans who were not white men like himself. He shunned pretense. He was comfortable in his own skin, and he had a special grace that enabled him to treat everyone, no matter what their status, as an equal. He was an adventurer, a father, a Christian true believer, and a Cold Warrior who pursued a breathtaking quest of exploration. And even years later, when his marine buzz cut had been replaced by a bald head with gray fringe, Americans still looked at him and saw a hero.

When President Kennedy decided to make his second and consummate argument for sending astronauts to the moon within the decade, he delivered an eloquent 1962 address at Rice University.[5] In it, he admitted that the challenge would demand much from Americans. Echoing his inaugural address's call to duty, he predicted that sacrifice would be required from the nation's people. "We choose to go to the moon in this decade and do the other things, not because they are easy, but because they are hard," he declared.[6] Kennedy placed space exploration in a historic continuum with the push to settle the American West. Voyaging to the moon became an extension of pioneering life, and John Glenn became a pathfinder on the way to the lunar surface.[7] "American exploration—from Lewis and Clark to the Apollo program—was acting both on a generic human impulse to seek knowledge and a deep-rooted American urge for inquiring, exploration, and the freedom of wide, open spaces," wrote space historian Asif A. Siddiqi.[8]

The symbolism defining the space program was not limited to frontier comparisons. "They were our gladiators in the contest with the Soviets," said Neufeld.[9] Observers compared Mercury astronauts to single-combat fighters, a centuries-old concept used in some cultures as a substitute for massive and bloody destruction. Each side of a conflict chose a champion—or hero—to fight in place of an army. "During the Cold War period small-scale competitions once again took on the magical aura of a 'testing of fate,' of a fateful prediction of what would inevitably happen if total nuclear war did take place," wrote Wolfe.[10] This symbolism gained power from the steeds on which the astronauts and cosmonauts rode—rockets designed as intercontinental ballistic missiles—weapons of nuclear war repurposed to serve as vehicles of exploration. Rather than being destructive, the rockets now became productive, opening the road to discovery.

From the start, the poised and emotionally open Glenn received more attention than his terse teammates. After he became the first American to orbit Earth, columnist George Dixon wrote, "An achievement like this requires a hero,"[11] and Glenn filled the part well. Another reporter predicted, "History may record [Glenn's Mercury flight] as a time when a creeping sense of national pessimism was halted, or, at least slowed down." He called it "the perfect union of a man, an event, and his time. There was a hole in the national ego, and John H. Glenn Jr. began to fill it with a remarkable personality."[12] A priest found another reason to rejoice: "It is heartening to have a hero of this caliber to take his place beside idols of the entertainment world who too often hold center stage for the young's admiration."[13]

Thousands of Americans, including many children, wrote to Glenn, describing the day of his flight as a highlight of their lives and comparing him to make-believe heroes like Superman. Just the sight of him "made them cry," Wolfe recalled years later, "and this made him a hero."[14] Some fans wrote songs about him; others composed poems. Like Achilles and Odysseus, he starred in a thrilling saga that touched the hearts of many in his native land and beyond. Millions had found a contemporary hero.

And Glenn's courageous acts did not begin in 1962 aboard his orbital craft, *Friendship 7*. As an uncelebrated marine pilot in both World War II and the Korean War, he flew a total of 149 combat missions. Especially in the skies above the two warring Koreas, he gained a reputation for being a hotshot, willing to sacrifice his own safety to protect the lives of allied troops and to defend American ideals. He understood personal danger and its place in the life of a military man. When asked about the perils of being an astronaut, he said, "Everybody is aware of the danger. . . . You feel it's important enough to take a risk."[15] Dale Butland, who served as Glenn's Senate press secretary and Ohio chief of staff, saw this attitude in the much older senator: "He was a patriot to his bones."[16]

During the years when the astronauts were becoming symbols of hope, the space age began to change the perspectives of earthbound Americans. As the 1960s unfolded, spaceflight's presence moved closer to the center of American life. From spaceman comic books to Tang, from Teflon to *Star Trek*, space-related phenomena invaded homes across the nation, while the space program laid the groundwork for much of the technology that defines life in the twenty-first century. Glenn and his colleagues became living icons representing America's new commitment, setting themselves apart from millions of Americans who would never volunteer. That's where the legend began.

A 1960 congressional report argued that even initial attempts at unmanned flight had made adult Americans think about "themselves, their country, and their world in broader, more knowledgeable terms." Children, the report concluded, were "even more deeply involved." Recognizing that some boys and girls pretended to be astronauts, the report argued, "The unique fact in the present situation is that never before have children rehearsed a role that really will not exist until they are adults."[17]

After he rocketed into America's first orbital flight, children across the nation dreamed of being John Glenn. On the day of Glenn's Mercury flight, one youngster rehearsing that role on his Texas school playground was John B. Charles, who later served as the project scientist on the age-related testing of Glenn during his 1998 shuttle flight.[18] Steve Robinson dressed up as John Glenn for Halloween as an eight-year-old, and later flew with Glenn on the shuttle *Discovery*. As payload commander, he gave orders to his lifelong hero. (Robinson does not believe he ever worked up the courage to tell Glenn about his childhood impersonation in a costume with "a lot of aluminum foil involved.")[19] Both Charles and Robinson found the real John Glenn to be every bit the man they had admired as children. Before meeting Glenn, Robinson had thought he was "a great person because he had done great things in his life." However, after knowing Glenn, Robinson realized "he had done great things because he was, at core, a great human being. It's not his achievements or his adventures that made that man great. It was just his personal nature and his values and the way he treated other people."[20] Charles recalled that those preparing for *Discovery*'s flight agreed: "He was probably one of the nicest people in America."[21]

The warm and positive Glenn fit neatly into the public perception of the space program at its beginning. "There is an emotional, intangible dimension to the human presence in space," a NASA report concluded.[22] The dream of reaching out and touching the moon undoubtedly is as old as mankind. Early humans surely noticed and wondered about the mysterious orb that traveled across the sky, sometimes casting light into the darkness of night.

For thousands of years, humans had been moored to Earth and its atmosphere by gravity, with no chance of escaping them. As a result, merely lofting a man into outer space represented a challenge when Kennedy announced the moon goal. And just as many Americans found the lunar objective dumbfounding, many NASA employees were startled by the magnitude of the venture. However, the bold nature of Kennedy's moon goal led people to imagine fewer limits. A common conversation starter at kitchen tables around the nation became, "If we can put a man on the moon, why can't we . . . ?"[23] If the young man in the White House could push his nation to the moon, why couldn't Americans solve problems closer to home? This sense of boundless potential made the population's avatars—the astronauts—seem unstoppable, unflappable, unforgettable.

Because astronauts worked on the front lines of the Cold War with the Soviet Union, Americans often characterized them as brave warriors. However, when called upon to speak, they often seemed like tongue-tied men who had stumbled into their wives' bridge parties—and that kind of behavior allowed them to serve as "everymen," at least to the white middle-class community.[24] In press conferences, the more relaxed Glenn seemed neighborly, like a man you would want living next door—someone who would play catch with your kids and teach them the difference between right and wrong. After Glenn's flight, Associated Press reporter Saul Pett reinforced this sense of Glenn as both a hero and an equal, saying, "He had done with surprising courage and skill what all of us Walter Mittys would love to have done."[25]

Throughout his ninety-five years, Glenn lived as his Presbyterian parents would have wanted him to live. He did not smoke or drink. He unashamedly loved only one woman from early adolescence until his death. He taught Sunday school, liked sports, and found peace in music. After drawing national attention, he consistently considered how his actions might affect young people like his teenage son and daughter. "John tries to behave as if every impressionable youngster in the country were watching him every moment of the day," a friend told a *Life* reporter in the early 1960s.[26] Glenn was a good man who tried to be better for the sake of the nation's youth. Some of his fellow astronauts thought his embrace of goodness was overdone. They had signed up for the challenge, not to serve as mentors to a generation of youngsters.

Astronaut Wally Schirra made no secret about his discomfort with Glenn's image consciousness. Nevertheless, he wrote, "John Glenn was America personified—baseball, hot dogs and apple pie."[27] Scott Carpenter, Glenn's one true friend among the seven Mercury astronauts, believed "the true John Glenn was more ambitious, more talented, funnier, and more charismatic than the humorless Calvinist of [Wolfe's] *The Right Stuff*." Carpenter called Glenn "a natural teacher" and marveled at his ability to speak to men, women, and children in the same way. Carpenter noted, "The Marine could carry on a conversation with a woman well beyond the cultural limits of place (the U.S. military) and time (the mid-twentieth century)."[28] Glenn's long and successful partnership with his beloved wife, Annie, may have contributed to that skill.

Fourteen years after his Mercury flight, Glenn used his winning personality, his proven sense of duty, and his well-known face to win his first of four consecutive terms in the US Senate from Ohio. "John was not a natural politician," Butland recalled, and he was not a prolific writer of laws, his legislative director Ron Grimes admitted.[29] Nonetheless, he had demonstrated nascent political skills even in the hierarchical US military establishment. As a senator, he worked hard, often tackling legislative tasks that no one else wanted to handle, such as the drudgery of improving governmental efficiency and the unsexy but vital job of riding herd on nuclear nonproliferation. He was never a Senate powerhouse, but he tried to use his office to improve life in the United States. Strong-willed and sometimes stubborn, he made his mark. "He was in constant motion," one longtime aide said.[30]

What Glenn wanted was to be a public servant more than a politician. Representing his constituents was more important to him than putting his name on groundbreaking legislation. Legislative aide John Haseley saw him as a man who "wasn't consumed by the Senate or by politics or by importance."[31] Unlike many traditional politicians,

Glenn "was collegial and wanted to run every investigation, every bill in as much of a bipartisan way as possible," Grimes said.[32]

In 1998, during Glenn's last year in the Senate and as he prepared to return to space at seventy-seven, the head of NASA saluted him at the National Archives. Daniel S. Goldin said it was appropriate to recognize Glenn at the institution that holds and preserves the nation's founding documents "because protecting, promoting, and providing inspiration for our freedom has been the course John Glenn has charted his whole life. He has charted the skies and then the stars and will do so again next week. Most of all, John Glenn has charted what it means to be an American and why the freedom that comes along with that is so very, very special."[33]

When he made his second spaceflight, aboard *Discovery*, Glenn seemed even more self-deprecating than he had been in 1962. As essentially an elderly test specimen, he could not say enough in praise of his six fellow astronauts. Space travel had changed a lot between his hours-long Project Mercury flight in a cramped capsule and his nine-day voyage aboard the comparatively spacious shuttle. Glenn jokingly told his *Discovery* colleagues that the shuttle was a lot more crowded than *Friendship 7*, but that was true only in terms of population.[34] In 1962 almost everything about Earth orbit and the effects of weightlessness was mysterious; in 1998 the mechanics of human orbital flight were well known, although much remained to be learned. Mercury astronauts shared military test pilot backgrounds, while shuttle crews included scientists in varying fields.

Looking at America's continuing celebration of John Glenn as a hero, it is impossible not to contemplate the role of heroes today and to wonder about American role models in the future. The First and Second Great Awakenings in eighteenth- and nineteenth-century American history were periods when evangelical Christianity drew passionate beliefs from the American people. Over the last half century, America has experienced what might be called a Great Yawning—a movement generated by conflicting forces and fired not by faith but by disbelief, cynicism, and a newfound and languorous numbness to the differences between right and wrong. Some people see this phenomenon beginning on a November day in Dallas when a loner named Lee Harvey Oswald deprived the world of one of its leaders, and a rocky 1963 weekend in which Jack Ruby proved that "an eye for an eye" is no guarantee of a fair trade. Americans lost an intelligent, youthful, inspirational leader, and the death of his misfit killer in no way leveled the scales of loss.

USA Today noted in 1998 that "the assassinations of the Kennedys and Martin Luther King Jr., Vietnam, racial riots, Watergate, the exposure of FBI and CIA abuses, oil shocks, all have led Americans to question their governments and heroes more." The editorial argued, "Good and evil have never appeared quite so unambiguous since."[35] Ben Wattenberg of the *Washington Times* wrote in 1999, "Vietnam, Watergate, O.J. [Simpson], and Monica [Lewinsky] have sapped our energy. It's so sad. America no longer does heroic things."[36]

Clearly, the change did not occur overnight but penetrated the American consciousness little by little until many Americans had lost hope for the future, trust in government, and belief in heroes. This new way of thinking has led some voters to follow blatantly dishonest politicians, shrugging off their false statements by accepting the premise that all politicians lie. Given our loss of heroes, neoconservative William Kristol has written, "We live in a lesser time. . . . Our America is in many ways a lesser America."[37]

Social scientists sometimes analyze cultures based on whom they choose as heroes. What can be said about a culture without heroes? Historian David McCullough has said that the need for heroes is "built into us." In his view, they "show us the possibilities of life and enlarge the experience of being human . . . and give us the capacity to endure the tragic."[38] Some of this change reflects the evolution of journalism. "The reporter used to gain status by dining with his subjects," said *New Yorker* writer Adam Gopnik. "Now he gains status by dining *on* them."[39] Historian Barbara Tuchman asked in the late 1980s, "Do we really have to know of some famous person that he wet his pants at age six and practiced oral sex at 60?"[40]

Somewhere along the way, heroism mistakenly became synonymous with perfection. JFK's dramatic rescue of his shipmates in World

War II, his deft handling of the Cuban Missile Crisis, his introduction of the first major civil rights legislation since Reconstruction, and his memorably eloquent speeches made many see him as a lost hero, but over the years, others have argued that his philandering makes him ineligible for the title. To some, civil rights leader Martin Luther King Jr.'s mistakes carry more weight than his great works, and Thomas Jefferson no longer stands high as the author of the Declaration of Independence when he is viewed simply as a prominent slaveholder who fathered enslaved children. Long before his relationship with Sally Hemmings could be confirmed, Jefferson had felt the sting of criticism. "None of us, no, not one, is perfect," he wrote. "And were we to love none who had imperfections, this world would be a desert for our love."[41] Today, many people overlook a simple fact: human heroes must be human and, therefore, imperfect. Did Glenn ever stumble? Yes, but not as a result of asking less of himself than he expected from others.

In the 1940s, World War II veterans, as a group, were treated as conquering heroes when they returned home and were later labeled the "Greatest Generation." The United States has fought four wars in the last fifty years, and not a single nationally recognized hero has emerged from any of them. The late senator John McCain may be the sole exception, although it is unclear whether his travails as a prisoner of war in Vietnam would have held a place in public memory if he had not subsequently become a US senator.[42] With an all-volunteer military, repeated assignments to war zones are more likely, but veterans often feel a lack of public support. In fact, the odds of a veteran between the ages of eighteen and thirty being homeless are twice as high as those for members of the same age group who have not served in the military. More than half of homeless veterans have disabilities and about half have mental health conditions.[43] Too many fight and return home to live in the streets without the thanks of a grateful nation.

The best-known Congressional Medal of Honor recipient in recent decades was a president of the United States who had fought more than one hundred years earlier: Theodore Roosevelt was honored in 2001 for his service in the late nineteenth-century Spanish-American

War. Roosevelt rode on horseback up and down the hills of Cuba while most soldiers around him were on the ground, where they could hide from enemy fire if necessary.[44] The United States honored him 103 years after his heroic service and 82 years after he died in his sleep in 1919. Roosevelt is often seen as an oversized boy at heart who, like Glenn, never abandoned his youthful curiosity about the world around him. Historians consistently rank Roosevelt among the nation's great presidents. He remains a hero to many despite some significant mistakes during his more than three decades in public life.

In ancient times, both Greek and Roman writers crowned heroes and acknowledged their flaws without portraying those weaknesses as roadblocks to important achievements. Many Americans jump at a single spurious allegation as a sign that a person is unworthy of admiration. And sadly, they often embrace unproven and untrue charges. For example, during Barack Obama's presidency, a notable minority of US citizens believed that Obama had not been born in the United States—and many clung to this idea years after Obama's birth certificate had shown that he was born in Hawaii. In 2016, Donald Trump, who had been a leader among the so-called birthers, finally conceded, "President Obama was born in the United States—period."[45] However, the baseless charge has never fully disappeared from public discourse.

In this cynical world, "we're starved for heroes," said Peter Gibbon, who has done extensive research about the absence of heroes in the information age and within a society that increasingly prioritizes individual achievement over what's best for society as a whole.[46] In a nation without heroes, how can we ever again face the kind of shockingly audacious challenge JFK threw before the nation when he announced the goal of putting a man on the moon in less than a decade? The United States never could have reached the moon without a cadre of men willing to die in the process—and the bravery of those men was fired, at least in part, by a nation willing to support their gamble, to celebrate their successes, and to mourn their losses.

After Glenn's Project Mercury flight, James Reston of the *New York Times* wrote, "The examples placed before a nation are vital. What we constantly observe, we tend to copy. What we admire and reward, we perpetuate. This is why John Glenn himself is almost as important as his flight into outer space, for he dramatized before the eyes of the whole nation the noblest qualities of the human spirit." He credited Glenn with "courage, modesty, quiet patriotism, love of family and religious faith."[47]

If, as Reston suggested, positive national role models like Glenn can affect private lives around the nation, it stands to reason that today's leaders who abandon dignity, embrace prejudice, and display rudeness set a bad example for the rest of the nation. Their coarse behavior seems likely to be replicated in boardrooms, classrooms, and family rooms across the United States. Instead of studying the big events of the past or envisioning bold endeavors in the future, many Americans today reside in a narcissistic and hero-free "selfie" culture. Demonstrating the courage to compromise is becoming a lost art, and the number of Americans who trust in government all or most of the time stood at 17 percent in 2019—down 60 percentage points since 1964.[48] Physical courage still survives and springs up in scattered events. Nevertheless, the courage to do the right thing is harder to find in public discourse.

Looking back at Glenn's first orbital flight, one observer noted, "Glenn cast a spell on the American people that never quite wore off."[49] When he died, almost every headline recognized him as a "hero." This book attempts to track the making of an enduring hero in hopes that he will not be our last. John Glenn was the kind of person America needs—a human being, neither perfect nor superhuman, but true to himself, his ideals, and his fellow citizens.

1

SMALL-TOWN BOY

John Glenn met his future wife, Annie, in a playpen at an age defined by skinned knees and sharp elbows. Both his parents and hers belonged to a Saturday night dinner club whose members dined together once a month. While members of the Twice Five Club shared good food, they corralled their kids together to simplify supervision. Although Annie was a year older than John, they became friends as little more than toddlers.[1] Throughout his life, Glenn never courted another female. Annie apparently dated another boy—once.[2]

The future astronaut and senator entered the world in Cambridge, Ohio, on July 18, 1921, and grew up in the village of New Concord, Ohio, about nine miles away. Later, in a B+ school essay, young John reported, "Soon after my arrival I was placed under the porch to see whether I was going to walk, crawl, or fly. They wanted to know what kind of an animal I would turn out to be."[3] In fact, he eventually walked, crawled, and, most famously, flew.

Whether he was hugging ropes to swing through the air à la Tarzan or frolicking in the water of local swimming holes like Tom

Sawyer, Glenn reveled in the experiences of being an all-American boy in small-town America. Fishing with safety pins as hooks and exploring the mysteries of his wooded surroundings, he embraced the adventures inherent in a happy childhood. Like all youngsters, he tested the boundaries of naughty and nice: he executed devilish Halloween pranks and solemnly absorbed Sunday school lessons. He was an intelligent, talkative, and playful red-haired boy with freckles to spare, and he enjoyed a feeling of freedom that was ripe with possibilities.

His father, John H. Glenn Sr., had married Clara Sproat shortly before boarding a ship for military service in World War I France. When he returned to Ohio, he worked as a fireman for the Baltimore and Ohio Railroad. However, he disliked spending many nights away from home, so he became an apprentice to a plumber. After he had learned his new trade, the family moved to New Concord when his son was two years old. After a short-lived plumbing partnership failed, the elder Glenn opened his own company, later broadening it to include a supply store.[4]

Clara Glenn, who had attended Muskingum College in New Concord for two years to earn teaching credentials, oversaw the store while her husband, who had only a fourth-grade education, pursued his craft. When the plumbing business stood on solid ground, the enterprising father became an associate Chevrolet dealer, which meant that he had a showroom with a couple of cars and worked in allegiance with a full-fledged Chevrolet dealer in nearby Zanesville.[5] He was a respected member of the community and an elder at the local United Presbyterian Church. Glenn's mother, too, became an elder, an unusual achievement for a woman of that era.[6]

New Concord was the kind of town where children felt safe to wander, but it also resembled the place envisioned by Robert Frost when he wrote, "Good fences make good neighbors." In the rural Ohio of the 1920s and 1930s, land represented stability; fences, respect. Inside the town, Muskingum College was the largest employer; on the outskirts, farmers erected fences and toiled within their bounds. On Saturdays, it was not unusual for villagers to see farmers driving horse-drawn buggies into town to shop.[7] Upon entering or leaving

John Glenn as a youth in New Concord, Ohio.
Courtesy of the Ohio Congressional Archives

New Concord, visitors saw a sign proudly proclaiming a population of 1,185,[8] enough people to fill less than 1 percent of the Superdome's seats.

Like most boys, young John Glenn sometimes found his way into trouble, but most of the time he toed the line his parents laid before him. Other than a profound sense of right and wrong, little set him apart from the village's other youthful residents. In fact, his childhood was quite ordinary for a smarter-than-average boy. His mother exercised her teaching skills as well as her love of poetry to train her four-year-old son to recite the centuries-old nursery rhyme "Little Boy Blue." In elementary school, he had the opportunity to skip a grade, but he chose to stay with his classmates and friends.[9] As a youngster, Glenn demonstrated a child's entanglement of financial precociousness and naivete: after his parents told him they could not afford the wagon he had picked out in a Columbus department store, he said, "Well, just give them a check and pay them later."[10]

Glenn spent his earliest years as an only child, but as he approached school age, his parents adopted a daughter, Jean. The decision to add to their family followed two miscarriages before John Jr.'s birth and one afterward. In his autobiography, Glenn said that he loved his little sister, but the difference in their ages made them unlikely playmates.[11]

Known to friends and family as "Bud," the young Glenn was no stranger to mischief. At around four years old, he earned his mother's wrath by exploring a nearby creek that she had clearly identified as a forbidden playground until he learned to swim. Terrified by what could have happened, she swatted his little legs with a switch yanked from a bush. Soon, welts arose, and she felt guilty. "My mother and dad didn't go in for paddling much," Glenn recalled. "Maybe a little swat on the tail with a flat hand once in a while, but not switching like that."[12] In second grade, he was one of two overly exuberant boys sentenced by their teacher to spend time in a dark closet. To the boys' delight, they found molding clay in their makeshift prison. When the teacher opened the door to free her captives, she was surprised to find a row of figurines.[13]

As a youngster, Glenn attended church and Sunday school regularly, and sometimes a Wednesday evening prayer meeting as well. Each year, the church dedicated one week to a church rally with special services, and he often participated. When he became older, he received an additional dose of religion through weekly meetings of the Hi-Y Club, a group sponsored by the YMCA. As a member of a fiercely religious family, he held strong beliefs. In the New Concord of Glenn's youth, Sundays were days of rest: no businesses were open except gas stations. Public service of alcohol was illegal on every day of the year.[14]

Even as a small boy at a time when airplanes were relatively new, he showed an interest in flying. He crafted toy planes from balsa wood. This hobby received special attention when epidemics closed schools and confined children indoors.[15] Lindbergh, who had crossed the Atlantic when Glenn was five, reigned as one of his childhood heroes.[16] As a youngster, Glenn flew for the first time in an open-cockpit biplane in Cambridge.[17] He was so small that he and his husky dad could share one seat in the plane's open cockpit.[18]

Full of energy, Glenn joined athletic teams that played mini-games of a few minutes during halftime at Muskingum College's games. Local newspapers, such as the *Barnesville Enterprise*, began reporting his athletic achievements as early as fifth grade.[19] In school, he competed for citizenship awards and entered essay contests. He enjoyed trumpet lessons and learned to play "Taps" at funerals with his dad.[20] The elder Glenn would play a line, and the boy would echo that line from some distance away.

During the Great Depression, the Glenns planted a large garden to minimize food bills, and once, young John was startled to hear his father and mother discussing the possibility that they might lose their house in a bank foreclosure. At eleven or twelve, the boy felt horror at the thought of losing his home. To avoid such an eventuality, John Sr. had built their home so that it was large enough to generate income by allowing the family to rent rooms to students and others. The National Housing Act of 1934, a part of President Franklin Roosevelt's New Deal, gave the Glenns and people like them a chance to acquire a loan with a longer term and a lower interest rate.[21]

Rudy deLeon, a personal friend of Glenn and one-time deputy secretary of defense, recalled Glenn's stories about life in New Concord during the Depression. US Route 40 went through the village, so it was not unusual for a hungry stranger to knock on the door seeking food. They were "migrating, looking for work, going east to west and west to east—they'd often come and knock on the door," deLeon said. "Whether it was just bread or something a little bit more elaborate," the Glenns tried to be sure "to have something when people came by."[22] Sharing with those in need was a habit Glenn learned from his parents.

One big thrill for Glenn and other New Concord children was helping the local train station manager by holding a long loop in the air as a train barreled through New Concord. The station manager placed instructions in the loop, which the engineer grabbed as he passed the station and dropped a few yards beyond the station. "You stood there by where the train was going to come thundering by about three feet away, and you held this up," he remembered with awe.[23] The Glenns' home sat close to the railroad tracks, and soot from passing trains sometimes came through the window screen and settled in John's bedroom, an exciting intrusion he didn't mind at all.

New Concord had no Boy Scout troop, so Glenn and some friends established the Ohio Rangers. Members embraced scouts' ideals as well as their yen for camping. Reiss Keck, Glenn's fifth-grade teacher, became their adviser. They produced a newspaper and fielded football and basketball teams. Each year they raised money by producing a play in a small auditorium. One summer they camped out for about seventy consecutive nights. On a single night, they took a break from the outdoor life to allow Annie's Girl Scout troop to use the camp. A gusty, torrential thunderstorm rousted the girls, toppling tents and washing away sawdust the Rangers had placed to make their camp more comfortable. Parents ran from their homes to rescue the girls on their first and last night of camping. During another summer, the boys chopped down trees and built a small log cabin to serve as their headquarters.[24]

Despite occasional misdeeds of his own, Glenn was at heart a rule follower and occasional enforcer. At one Ohio Rangers gathering, C. Edwin Houk sang "Hail, Hail, the Gang's All Here," including the line "What the hell do we care?" An outraged Glenn condemned Houk's use of profanity and told him to stop. "I think he was ready to knock my block off," recalled Houk, who grew up to be a minister.[25] Houk remembered another occasion when several teens were riding in Glenn's car, and someone commented on a young mother pushing a baby carriage, calling her "a good-lookin' gal." Glenn responded negatively, exclaiming, "Hey, that's a married woman!"[26]

The New Concord community provided good foundations for a life shaped by patriotism and piety, Houk later wrote. "Many stones went into those foundations," he said. "John and I learned friendliness from New Concord's people. As members of the town band, we marched in many a parade, playing tunes like 'America,' and 'Battle Hymn [of the Republic],' learning a kind of patriotism that is growing scarce today," he recalled in 1962. They also "got a good grounding in the scriptures."[27]

Throughout his youth, Glenn learned that the best way to obtain a yearned-for possession was to earn money himself. He traveled the village's streets selling ten-cent bundles of rhubarb from the family garden and twenty-five-cent packages of sassafras from the woods near their home. Young John sometimes worked for his father, standing outside a building where plumbing work was underway. In response to a friendly yell from his dad, he would fetch and fashion whatever pipes were needed.[28] The harsh winters of 1936 and '37 also provided opportunities for hardy youths to earn money shoveling snow. For several years, he delivered the *Columbus Citizen* and washed cars to accumulate extra cash. He used his earnings to buy a used bike for sixteen dollars.[29] Beginning in high school, he spent much of three summers working at a YMCA state camp—Camp Nelson Dodd. The first summer, he oversaw a serving crew in the kitchen. In later years, he drove a truck and offered counseling to campers.[30]

In junior high school, John and Annie had officially become a couple, and the relationship continued unbroken from that point onward for more than eighty years. Annie was an 85 percent stutterer, which meant that her speech halted on 85 percent of the words she spoke. Her father stuttered, and as a child, "I didn't realize I was a stutterer," she later remembered. Like many revelations, this one arrived in sixth grade on the doorstep of adolescence: "I got up to recite a poem and I evidently was having a really awful time and one of my classmates laughed, and that was my first realization that I was not like all of the others."[31] Glenn recalled that her stuttering had never bothered him. He did notice that in high school "she never was in a school play and was never asked to give a recitation in school."[32] As she got older, she developed a coping mechanism to use when shopping: writing notes for sales clerks to explain what she wanted to buy.[33]

As an adolescent, Glenn loved Halloween and the previous night, "Corn Night," when kids threw kernels of shelled corn onto neighbors' houses, leaving a mess for grouchy grownups to clean up in the morning. He often participated in harmless pranks, such as moving a neighbor's lawn furniture to another part of the mile-wide village. However, he learned a harsh lesson one Halloween when his mother awoke him at 2:00 AM and told him that Town Constable Mike Cox was waiting downstairs for him. A sign painter had complained to the constable that his sign had been taken, and he could not find it. He had seen Glenn among the ne'er-do-wells who had swiped it. Now fully awake, the terrified boy confessed everything and revealed exactly where the sign could be located. The constable left happily without the threat of punishment.[34]

During his teenaged years, Glenn received an invitation from Reiss Keck to join his sister and him on a drive through northern Ohio to the Saint Lawrence River and up to Montreal, before beginning the return swing through New England and New York City. Ever the teacher, Keck stopped in towns and convinced plant managers to offer them tours of factories and power plants. Young John insisted on jumping into the ice-filled Saint Lawrence River and enjoyed the threesome's climb of Mount Washington in New Hampshire. In the years before this trip, he had left Ohio only twice, once when his father took him to the Chicago World's Fair in 1933, and much earlier when the family had visited his mother's sister in Washington, a trip he barely recalled.[35]

In high school, he embraced the ideal of public service as taught in civics class by teacher and later principal Harford Steele. Glenn learned a great deal about history and government from the inspiring Steele, and he wrote a research paper on the US Senate.[36] A capable athlete, though not a great one, he lettered in basketball, football, and tennis. Basketball was his favorite. He regretted taking the position of center in football because he disliked the inevitable battering he experienced in every game. Moreover, he enjoyed taking part in school plays and served as president of the junior class.[37] While John and Annie watched other friends break off relationships and embrace new ones, "we were just an entity," he remembered.[38]

After graduation, he accompanied three friends on a trip along the Saint Lawrence River, south through New England, and on to the 1939 World's Fair in New York. The boys camped on Long Island while enjoying the exposition, which was themed "The World of Tomorrow." After that, he returned to Camp Nelson Dodd for his final summer there.[39]

As he looked ahead, Muskingum College was a logical next step. Annie already was a sophomore there when Glenn joined the freshman class. The college, founded in 1837, was affiliated with the United Presbyterian Church. As a young adult, Glenn remained deeply religious. Houk recalled a discussion in a college Bible class when another student questioned the scriptures' believability in describing the part-ing of the Red Sea. Glenn raised his hand in the air and said, "It's in the Bible, isn't it?" When someone confirmed that it was, he said, "Then that should be enough"[40]—end of discussion, as far as Glenn was concerned.

To fulfill the work requirements of his college scholarship, Glenn served as part of the college maintenance staff. While he was at Muskingum, Glenn thought he might become a physician or a chemist, but the dream of flying never left him. In 1940 and 1941, as the United States braced for possible entry into World War II, Glenn enjoyed an opportunity to participate in the government-financed Civilian Pilot Training Program. He received course credits for training three after-noons a week in New Philadelphia, which was an hour's drive away. He earned his private pilot's license, and afterward, he saved money to rent planes for jaunts around Ohio.[41]

By this time, John and Annie had been a couple for more than a third of their young lives, but they were not openly affectionate, even in college. "They never held hands or anything like that in public," a classmate remembered. "They weren't the mooning kind. But they were always together—swimming in the college lake, picnicking, play-ing tennis, ice-skating, and going on hay rides. Everyone just took it for granted that eventually they would get married." Once, during a lecture, a professor asserted that young people should avoid "going steady." Glenn stood and asked, "Wouldn't you agree that it was all right as long as you knew you had selected the right person?" His class-mates laughed knowingly.[42] His heart belonged to Annie. Neverthe-less, for one of his classes, Glenn wrote an essay called "The Choice of a Mate," in which he admitted, "We can no longer believe that there is only one person who can be a real mate."[43] He failed to add his firm belief, however, that he had found a perfect match.

John Glenn never relinquished his roots. "He was a man of New Concord, Ohio," deLeon said. "He's not a person of San Francisco or New York City, but he really came from the heartland. And it was those values that he took into the Marine Corps for World War II and Korea, and then into Project Mercury."[44]

2

OFF TO WAR

December 7, 1941, rerouted the lives of millions of young men like John Glenn. When an angry swarm of Japanese planes breached the skies above the US base at Pearl Harbor in Hawaii, they carried out a devastating surprise attack, damaging or destroying twenty US naval vessels and taking more than twenty-four hundred American lives. Shock claimed a stranglehold on a nation that, until that calamitous morning, had been home to many isolationists. The following day, President Franklin D. Roosevelt requested and received a declaration of war against Japan. Three days later, Japan's European allies—Germany and Italy—declared war on the United States. Suddenly, the nation faced war on multiple fronts.

Glenn's life would begin to expand, making way for first-time experiences, such as comradeship with men from all parts of the United States, voyages to the other side of the world, and combat missions. No longer cocooned in New Concord, he would experience the meaning of courage, the diversity of American life, and the pain of losing a buddy in a moment when he looked away.

The seeds of the man he would become lay in Glenn's World War II experiences. The heroic image of the World War II soldier became the model for the Cold War space warrior: a tough, self-sacrificing crusader for good in its holy war against evil. Both struggles seemed inevitable, inescapable, no matter the cost.

War is deadly and destructive, but it is also a factory of change. Nothing can belittle the loss of more than four hundred thousand American lives and millions more in other nations during World War II, but through a complex conglomeration of factors, including alliances, late entry into the war, scientific breakthroughs, and geography, World War II provided an incredible boost for the United States' position in the world. Efforts to win the war ignited a transformation in which American industries boomed, US technology excelled, and the nation rose to global leadership.

American ingenuity played a role in this triumph, but the nation's biggest asset was a simple one: the United States emerged whole from the war, oceans away from the bombed-out cities, the blood-drenched battlefields, and loathsome memories of the war's haunting horrors—brutal concentration camps, sex slavery, mass murders, gas chambers, human ovens, the devastating effects of atomic bombs in the Pacific theater, and tattered yellow stars among the rubble of Europe. Americans had their own shames to acknowledge over the coming years: the imprisonment of Japanese and Japanese American civilians and a much smaller number of German Americans, the nation's slowness to help Jews emigrate from Europe, and racial segregation inside the military and much of the nation. However, in the immediate wake of the war, these misguided policies failed to sully the sense of triumph in a historical moment celebrated as "the good war," a conflagration in which the United States and its allies crushed oppressive regimes. More recent American wars have not carried the unthinkable monstrosity, the unspeakable dread, or the unending historical romance of World War II. The mythology of a war seen as a clear confrontation between right and wrong would transform the victors into heroes against whom future generations would be measured.

After Pearl Harbor, Glenn quickly decided to forsake college to become a military pilot. The day of the attack coincided with Annie's senior organ recital. Rather than spoiling her big day, Glenn waited until her fingers had struck the final notes before sharing news of the attack. Annie quickly decided that she would marry him and would go on to accompany him to his military postings in the United States, relinquishing a scholarship to study at Juilliard in the process.[1]

Hoping to be deployed expeditiously, Glenn stopped his coursework at Muskingum as 1941's fall semester concluded. He and a friend, Dane Handschy, reported to the US Army Air Corps office in nearby Zanesville, Ohio, and signed up. After being sworn in, they were told by a recruiter to await orders. To their surprise, weeks passed without any word from the army. While they waited, the duo sought ways to fill their time. A farm owner hired them to do custom plowing for farmers who did not own tractors. Seeing the job as a way to build a nest egg and occupy their hours, both went to work driving tractors and sometimes labored until midnight, using headlights on the tractors to light the way.[2] They tilled hundreds of acres in a few weeks. Neither minded the work, but they longed to soar high above the rich soil of the United States as part of the nation's armed forces.

When March arrived without any word from the army, Glenn returned to Zanesville, this time visiting the navy recruiting station with news that he had sixty hours of flight experience. He was not entirely sure where he stood within the army's mighty bureaucracy, but he wanted to start military training. Again, he took an oath, and his navy orders arrived within two weeks. He and Annie had hoped to marry before he reported for duty, but navy regulations denied flight training to married men, so they settled for engagement. "Didn't have any money, but we got a ring," Glenn recalled. It cost $125 or $150. They planned to marry when he earned his wings.[3]

Training, which began in May 1942, would take Glenn to parts of the nation he had never seen—from Iowa City, Iowa, to Olathe, Kansas, and finally to Corpus Christi, Texas. At each base, the work was demanding, and training offered new challenges. He said goodbye to Annie and his family at the New Concord train station and rolled through the night toward a new life. "Leaving there under those

circumstances with everybody concerned that I was off to war as my
dad had done before was quite an emotional thing," he remembered
half a century later. "Even changing trains in a big station like Chicago
was a big thing for me," he said."[4] At the Iowa City station, guides
rounded up servicemen arriving for the Naval Aviation Pre-Flight
School, which was located on the University of Iowa campus, where
he spent three months. There, he experienced the brutal and demanding physical challenges of boot camp as well as the necessary classroom
preparation to become a naval officer and to prepare for training as a
military pilot. In addition to navy regulations, he studied engineering,
aerodynamics, and navigation.

Heading the physical training program was University of Minnesota football coach Bernie Bierman, who had guided his team to
five national titles by the time he joined the navy's training program.
To augment grueling military marches, Bierman established practice
in sports ranging from gymnastics to swimming. He attracted All-
American athletes to be navy trainers and assembled a formidable
football team. In one game, Glenn faced off against an All-American
center. "I don't believe I ever got more beat up in my life," he lamented.
"I wasn't sure I quite understood why getting beat up like that would
help me be a good pilot, but that was part of the training anyway."[5]

As Glenn began to rub shoulders with young men from other
regions of the United States for the first time, he felt that he "came
into World War II with a very decided view of patriotism and sense of
right and wrong that maybe was a little more strict or a little more pro-
nounced than many people had from their backgrounds."[6] His parents'
devotion to the Presbyterian Church and his small-town upbringing
had not prepared him for the rampant swearing and lascivious comments that often circulated in the uncensored, all-male atmosphere of
a World War II military training camp.

At the end of his time in Iowa, his "Officer Aptitude Report
for Student Officers of Cadets" showed an overall score of 3.7 on a
4-point scale—above average but not outstanding. He received grades
on intelligence, judgment, initiative, force, cooperation, loyalty, per-
severance, endurance, industry, and military bearing. No record of

bad behavior appeared on his record, and he had demonstrated no noticeable weaknesses. His evaluator scribbled a simple note: "Very fine character—Handles assignments very efficiently."[7]

Next stop: Olathe, Kansas, where Glenn finally got to climb into a plane along with an instructor. Before arriving in Olathe, he traveled to Dayton, Ohio, where Annie was working as a secretary for the army air corps.[8] He stayed in the YMCA for a few days, and then headed to his next assignment. Everything at Naval Air Station Olathe was new, from the barracks to the airfields. Glenn joined the station's first training class. He began pilot training in an open-cockpit Stearman biplane. Though familiar in some ways, it was significantly different from the plane on which he had earned his civilian license. Stronger, faster, and larger, it could cover longer distances. The Stearman's airborne strengths invited daredevil stunts, such as flying upside down and performing breathtaking loops. Glenn enjoyed this training and felt comfortable with his ability to fly, as well as his mechanical understanding of the aircraft. "I'd grown up being sort of mechanically inclined around my dad's cars, and I'd taken engines apart and put them back together," he recalled. Still, in Olathe there were new lessons to absorb, such as how to use ordnance and how to make flight prechecks. After he had soloed, he began learning to fly in formation and at night. He moved to Corpus Christi, Texas, his last training station, in early November 1942.[9]

During the final phase of his piloting education, he became proficient in flying heavier planes. The most unnerving part of his Corpus Christi training was being forced to guide a plane by instruments alone. With his view obstructed, moving in three dimensions became something new and challenging. At that point, he chose to become a marine pilot rather than a navy pilot because he admired the marines' historic battle record. He made one big mistake in Corpus Christi: believing a rumor that marines soon would be flying P-38 fighters, he thought he needed experience flying a multiengine plane. To prepare, he and a friend, Tom Miller, decided they should seek training in the only multiengine planes at Corpus Christi: PBY flying boats.[10] They invested significant time in mastering the PBY—a decision that would come back to haunt both men. At the end of training, Glenn graduated, making him one of more than thirty-five thousand men to earn his wings at Naval Air Station Corpus Christi during World War II.

At that point, Glenn had one immediate goal—marrying Annie. He dashed back to New Concord, where they wed in a simple ceremony at the United Presbyterian Church on Tuesday, April 6, 1943. After a two-day honeymoon in Columbus, Ohio, they returned to New Concord briefly before heading to his next assignment in Cherry Point, North Carolina.

The drive to North Carolina was arduous. Their 1934 Chevy Coupe experienced repeated flat tires, and in a nation with rubber rationing, this became a bedeviling problem. When he finally reached Cherry Point, the would-be fighter pilot experienced a shock: he expected to see rows upon rows of P-38s, but there were none. The rumor had been wrong. He was assigned to fly a North American B-25 Mitchell, an established medium bomber. Compounding his frustration, there were too many pilots and not enough planes on the base. Consequently, the fliers spent most of their time in classrooms, not in the air. After just three weeks, both Glenn and Miller received reassignment to Fort Kearney in California. The Glenns drove across the country, stopping at the Grand Canyon for half a day before reaching their destination.

When they arrived in San Diego, Glenn and Miller were horrified to learn that they had been assigned to fly bulky DC-3s in a utility training squadron; their decision to get multiengine experience, which had seemed like a good strategy, had led to an assignment flying what Glenn called "boxcars" instead of sleek fighters. Desperate, the two young fliers visited a marine F4F Wildcat fighter squadron also at Fort Kearney. They met the squadron leader and asked whether he had openings for two pilots. He said he would welcome them, so they returned to the transport squadron to find out how to arrange a transfer. Their commanding officer was outraged that these two rookies had sought reassignment without consulting him. He ordered them to stand at attention and "chewed us out royally," said Glenn, who called the officer's tirade the "worst dressing down" he ever received in a twenty-three-year Marine Corps career. "I thought we'd never get transferred," he said, "and lo and behold, a few days later, our orders came through, and they transferred us over to the fighter squadron."[11] Within days of reassignment, Glenn, Miller, and the entire fighter squadron were transferred to El Centro in California's arid Imperial Valley.

The summer heat at the air station was abominable, so pilots tried to avoid flying during the scorching hours of midday. The Glenns moved into a hotel room without air conditioning as the temperature sometimes skyrocketed to 125 degrees.[12] Some marines playfully fried eggs on the ramp at the airfield.[13] Instead of following regulations and dumping all leftover fuel, pilots routinely used aircraft fuel to partially fill their cars' gas tanks. This bit of malfeasance allowed them to avoid the full effects of gasoline rationing.[14] Meat rationing was another issue. The pilots had meat rationing chits, but Glenn quickly discovered that the easiest way to get a juicy steak was to cross the border to the Mexican city of Mexicali, where there was no rationing.

A rush to the altar was as much a part of World War II military life as rationing. One Sunday night at 10:00 PM, Glenn received a call from a friend, Henry Knauth, who wanted the Glenns to drive him, fellow pilot Stan Lutton, and two young women to Arizona, where the pilots could marry their girlfriends. They had met their brides the previous weekend. The Glenns drove both couples to Yuma, where they spotted a sign for a justice of the peace. Arriving after midnight, they were greeted by a man in a bathrobe. The same man, still wearing his bathrobe, performed a double wedding ceremony.[15]

Before they could train in real fighters, Glenn and Miller had to fly fifty hours in SNJs—retractable-gear planes. After that, they climbed aboard their first genuine fighters, the somewhat outdated F4F Wildcats. The planes seemed slow, and like the SNJs, they required pilots to crank the landing gear up and down by hand. "It had a very narrow landing gear, and it had a tendency to want to veer off the runway a little bit when you were landing, if you got the tail down too soon. There was trouble controlling it in a crosswind, and so it was not unusual at all to have people run off the runway and be bouncing out across the sand there at El Centro," Glenn said.[16] At this air station, he wrote in a briefly kept diary, "We flew all types of section and division tactics, gunnery, navigation hops, strafing and worked on ground-air problems."[17]

In September 1943 new F4U Corsairs arrived, to Glenn's delight. He liked the planes, but he said, "The air, not the runway, was the Corsair's element." In other words, they flew well if handled carefully, but landings were troublesome. Once, he was making a fast dive and found himself being thrown about the cockpit violently before he could regain control. He resolved that he would not go that fast again unless a Japanese fighter was "on my tail and I wanted to shake him off."[18] The Corsair had "speed, climb, power, and firepower that makes it the best fighter that is in combat today," concluded Glenn, who won a promotion to first lieutenant in October 1943.[19]

Vought, the Corsair's manufacturer, hired Glenn's childhood hero, Charles Lindbergh, to visit air bases and fly the Corsair. Lindbergh arrived at El Centro in October to fly with the squadron and talk about the Corsair. Almost six years old at the time of Lindbergh's flight, Glenn remembered it as a celebrated moment during his childhood. "He was a legend and a pioneer and every pilot's hero," Glenn said.[20] "I still remember [Lindbergh's flight] as being an area of great interest around the community, more so than the Wright Brothers, as far as influencing me," he told one interviewer.[21] After Lindbergh's 1940 support for the isolationist America First Committee, his shining image had been tarnished, and in the following year, he had resigned his commission in the US Army Air Corps when President Franklin Roosevelt criticized his political position. In the wake of Pearl Harbor, Lindbergh supported the war effort. Promoting the Corsair as a warplane was a way of applying fresh veneer to his image.

As Glenn was finishing his training, several of his buddies had departed for war assignments. Before F. Lee Lemly II left for deployment at Funafuti in the Ellice Islands, he and his wife shared their last weekend with the Glenns. Shortly after reporting for duty in the Pacific, Lemly died mysteriously. He bailed out over the Pacific, and rescuers found his chute but no sign of Lemly. He had married only months before entering the war.[22]

By the time Glenn was ready to leave El Centro, he had reached several conclusions about being a fighter pilot: "The high speeds and fast maneuvering of a fighter plane make it necessary to react instead of think over a situation. . . . You do not see that the plane is changing attitude, but rather feel it first." He believed that this experience builds "absolute confidence in yourself and your plane that makes a fighter pilot believe that there actually is not another pilot and plane that can whip him in all the world."[23] As Glenn's assurance solidified at El Centro, Annie, too, got a chance to use her talents. She served as organist at a Methodist church.

Glenn last flew at El Centro on January 6, 1944. Almost exactly a month later, he headed to his first wartime assignment. When he and Annie parted in San Diego, he said, "I'm just going down to the corner store to get a pack of gum," and she replied, "Don't be long."[24] This became a ritual farewell whenever he faced a dangerous assignment that would separate him from Annie, whether on a marine assignment or aboard a rocket to Earth orbit. Having parted from his wife, he boarded the USS *Santa Monica*, a onetime banana boat, that carried troops toward Hawaii and the war in the Pacific. Glenn, who had never traveled at sea, suffered severe motion sickness. "The first day at sea I didn't care a whole lot whether I lived or died or fell overboard," he recalled.[25] Ultimately, he slept on the deck so that he was in the open air rather than in a stuffy cabin and close to the rail in case nausea overcame him.[26] While he struggled, Annie drove back to New Concord, where she planned to stay with her parents until his return.

Glenn reached Hawaii on February 12. His squadron's planes had not arrived, so he and his comrades "had a lazy man's paradise," he said. They spent hours lounging by a swimming pool and watching movies. He rented a surfboard, but "I never did get to the point where I could do much with it except paddle around." At Waikiki Beach, he bought Annie a hula outfit. On February 21, he and two other pilots joined twenty enlisted men on a twelve-hundred-mile flight to Midway for more training.[27] Their ultimate destination was the Marshall Islands.[28]

By the time Glenn arrived, Midway already had played an important part in the American war in the Pacific. In mid-1942, US forces won a major victory there. Japanese admiral Yamamoto Isoroku had planned to capture Midway and to ambush the only US aircraft carriers still in the Pacific after the Pearl Harbor attack. Yamamoto did not know that US intelligence already had decrypted Japanese fleet codes. What the Japanese had in mind was to distract US forces with a decoy movement. Admiral Chester W. Nimitz knew what was coming, and as a result, he didn't take the bait and was able to turn the tables on Yamamoto. US scout planes spotted the Japanese forces early on June 4, and Nimitz seized the element of surprise from Yamamoto. US forces won a costly but rewarding triumph: American forces lost thirty-six of forty-two torpedo bombers but sank *Akagi, Kaga, Soryu,* and *Hiryu*—Japanese carriers with 322 aircraft and more than five thousand men aboard. The loss contributed to a decline in Japan's air war.[29]

Arriving at Midway more than a year after the battle, Glenn found a peaceful spot blessed with beautiful sandy beaches. The Midway Atoll, about five miles in diameter, is part of the Hawaiian archipelago and encompasses several small islands, including Sand Island, where Glenn's squadron was based. A nature lover, Glenn was fascinated by resident "gooney birds," a type of albatross. He enjoyed watching a "peculiar dance" in which they faced each other and began "bobbing, weaving, and clicking bills in a very elaborate routine."[30]

After setting down shallow roots in Midway, the pilots got to work. At all times, four planes had to be prepared to take flight within two minutes to meet the threat of Japanese attacks on the US submarine base at the atoll. The pilots spent a lot of time in the air, honing their skills and testing the limits of their aircraft. They received additional training in navigation and gunnery.

Squadron members lived in Quonset huts constructed from prefabricated materials. One of the group's favorite pastimes was listening to Tokyo Rose on the radio. Sometimes, the Japanese propaganda maven offered only party-line chatter, but at other times, there were startling hints of inside knowledge: On the day after Seabees (members of the US Naval Construction Battalions) finished building a runway at Midway, she congratulated them on their success.[31] At times, the pilots experimented with spear fishing, and Glenn managed to pierce an octopus, which startled him by grabbing the spear with all eight of its tentacles. While swimming one day, he encountered a large sea turtle, and the Ohio boy was relieved when it passed him without incident.[32]

Sometimes, members of marine air squads and submarine crews traded adventures. Pilots took submariners up in two-seater planes for aerobatic rides, and on one day Glenn and other marines talked their way onto the USS *Barb* for a practice submarine cruise. The *Barb* made about ten dives while simulating war conditions.

> When we hit our 300-foot depth, some of the packing around the periscope blew, and in the little control room where we were, it was like someone turned a fire hose loose in there, hitting everybody, and it was soaking. Of course, those of us that didn't know much about submarines thought we were dying at sea for sure, and I certainly did. The submariners were getting a big kick out of it because they knew what we were not aware of, was that the water was being pumped out the bottom just as fast as it was coming in the top, and this [equipment problem] meant that they were going to get to stay at Midway for another couple of weeks before they went out on patrol.[33]

(The *Barb* later made history by entering Tokyo Harbor and resting on the bottom until a new aircraft carrier was launched. The US sub sank the carrier immediately and escaped. The *Barb* was able to infiltrate and escape the harbor at times when submarine nets were opened to allow passage by Japanese ships.)[34]

Despite pleasant diversions in their exotic new home, the war and its dangers were never far from the pilots' minds. On Easter Sunday one flier, known as Drifty, started down the runway at high speed just as another pilot, Joe Johnson, inexplicably pulled in front of him. The two planes collided. Drifty was able to achieve some altitude to minimize the impact on his plane, but Johnson apparently died instantly. Drifty ended up spinning around on the runway before bringing his plane to a stop. A PT boat buried Johnson at sea. "Joe was one of the best-liked men in the squadron and was usually the 'life of the party' wherever he went," Glenn wrote in his diary.[35] Thinking back on that tragedy, Glenn said, "If you're a combat pilot and you start dwelling too much on mortality and all that in war, you probably wouldn't fly the way you have to do to help win the war. So that wasn't something that I mooned over or thought about excessively."[36]

Johnson was the only pilot to die on Midway during Glenn's deployment there. Others came close. While landing, one fighter's wing hit a sand dune; both wings were torn off, and the plane hit the ground so hard that the engine flew off. The pilot found himself on the ground, backward and upside down. Miraculously, his only injury was sand in his eyes.[37] Another pilot was flying at high altitude when he lost control of his plane. He bailed out and was found safe in his raft.[38]

Glenn and his cohort left Midway June 11 and returned to Honolulu. They were to get twenty-two new F4U Corsairs, fresh from the factory. While they waited, the marines settled into the Moana Hotel. After a week, the command officers and a few others flew to Majuro in the Marshall Islands. Glenn and most of his fellow pilots stayed in Hawaii, he said. "A usual day's activity during this time was to wake up about 0900, yawn, stretch, and go back to sleep until 1030, when we would again awake and this time, we would muster enough energy to go down and lay on the beach for a while," he recalled. They dined on sandwiches from a lunch stand, and when they returned to their rooms, someone would say, "Call the boy for a half-dozen Cokes," Glenn said. "I must have had a hundred Cokes while we were there."[39] Finally, on June 28, the pilots boarded the USS *Makin Island* headed for Majuro. The ship was a lightweight escort carrier, later sardonically nicknamed a "Kaiser Coffin." These ships rode high in the water and tended to roll, but Glenn experienced no motion sickness on this voyage and arrived at Majuro on July 3.[40]

Glenn immediately noticed differences between Midway and the Marshall Islands. His new station was warmer and more tropical. Almost daily afternoon rains led pilots at Majuro to favor flying early in the morning. Once, a whole squadron had become lost in a storm.

John Glenn as a fighter pilot during World War II. *US Marine Corps*

The pilots landed at sea off an island and lost several men while trying to get ashore across a coral reef.[41]

Like Midway, the Marshall Islands already had been the scene of a major battle. Operation Flintlock was the American strategy to capture the Marshalls and fortify the US position in the Central Pacific. The campaign began in January 1944 after US forces successfully took control of the Gilbert Islands in late 1943 following the battles of Tarawa and Makin. In early 1944, US forces seized most but not all of the Marshall Islands.

On day one in Majuro, Glenn experienced something totally new: having his plane thrust into the air by a catapult, a method of taking flight used when adequate runways were not available. "That is really a boot in the tail," he wrote. "You sit there, get everything all ready and check the plane to see that everything is all set, sit way back in the seat with your head against the headrest, rev up the engine to about full power, reach over and give the signal, and about one second later you are sitting out in space flying and not too sure about how you got there." In training, as soon as a pilot was catapulted, he just swung around the island and landed so that he could ride the catapult again.[42]

One of the most dramatic events in Glenn's World War II service occurred during his first combat mission. The squadron's assignment was to fly about one hundred miles to the island of Taroa and to strafe the ground, opening the way for dive-bombers flying behind them. Traveling at treetop level, the fighters' goal was to suppress antiaircraft fire that might hit the bombers. Glenn remembered looking over at his wingman, Monty Goodman, flying beside and slightly behind him. As they approached the target, he saw Goodman drop back a bit as he should when they neared the island. Glenn then reviewed the run's most vital targets and made the attack. He never saw Goodman again.

After the bombing run, the squad rose above cloud cover for a rendezvous, and the only fighter missing was Goodman's. The squadron commander as well as Glenn and two others remained in the area, trying to determine what had happened. One dive-bomber pilot reported seeing a plane splash into the water. Another pilot saw something streaming out of Goodman's left wing. After a brief search, the team discovered an oil slick about a mile and a half off the island. A thorough search revealed no sign of Goodman himself.

Glenn and Goodman had been close. They had sailed together on the *Santa Monica*, socialized in Hawaii, and served together on Midway. At one point, Goodman had sprained an ankle playing volleyball, and Glenn had carried him on his back across hangar decks.[43] When

Glenn returned to base after Goodman's disappearance, he started to stow his parachute in his locker when he noticed Goodman's name on a nearby locker. "The pent-up feelings of the last few hours let loose, and I stood there and sobbed. . . . The war of waving flags and musical send-offs was gone, and I had seen my first day of the real thing. I had lost one of my closest friends, and war had suddenly become very, very personal," Glenn wrote in his memoir.[44]

Despite losses, flying in the war zone became routine quickly. Glenn flew almost daily, and sometimes twice a day on patrol or on attack missions. The twenty-four-plane squadron's targets often were the four Marshall Islands not under US control. There were Japanese on these islands with antiaircraft weapons but no supply line from Japanese military forces. Marine fighter squadrons maintained constant pressure to guarantee that the Japanese could not re-arm the islands and use them as bases for new attacks.

One of the squadron's crucial missions was an attack on Nauru, about three hundred to four hundred miles away from Majuro. This represented Glenn's longest open-ocean flight. Nauru was a great source of phosphates, and the US military believed the Japanese desperately needed the island to make fertilizer, which was necessary to raise crops to feed the population at home.[45] Army, navy, and marine planes joined in a continuing effort to end Japanese use of Nauru, previously under British control.[46] Glenn's squadron first flew to Tarawa, where the fliers waited for bad weather to clear. Because there were no islands between Tarawa and Nauru, the squadron had to use dead reckoning to reach its target. A twin-engine craft handled navigation and led the other planes in formation until pilots sighted the island. Antiaircraft fire from the surface was extremely heavy, but the squadron returned with no pilots lost.[47]

On the job, Glenn gained a reputation for precision flying. Marine lieutenant colonel John Masson said, "Johnny would fly up alongside you and slip his wing right under yours, then tap it gently against your wingtip. I've never seen such a smooth pilot."[48] Glenn was proud to lead a Jaluit Island firebombing mission of about five hundred nautical miles round trip. His team bombed several gun emplacements, with some carrying single two-thousand-pound bombs rather than the typical three one-thousand-pounders.[49] In addition to bombs, their planes were armed with six .50-caliber machine guns, which were accurate up to about twelve hundred feet.[50] There were no bombsights in the aircraft to improve the pilots' accuracy, so they had to train intensively for bombing runs.

During Glenn's missions, "I was hit, I think it was five times," he said. "Three of the times I was hit were single or multiple bullet holes. One time, however, I was hit in the leading edge of the left wing with something that probably was the Japanese equivalent of our twenty-millimeter because it blew a hole . . . that was about the size of my head. . . . I remember I landed a little bit fast to make sure I had good control. Another one was a little bigger hole."[51]

While Glenn was on Majuro, Lindbergh visited the squadron. The aviator wanted to be part of the fight, so he was operating as a volunteer civilian adviser and went into combat about fifty times in the Central Pacific, including several missions with Glenn's squadron. He devised a way to carry bombs as far as Japan without refueling,[52] and he found solutions to landing gear and wing tank problems that were plaguing the Corsairs.[53]

Glenn and his comrades made some of the first napalm strikes in the Pacific. An extremely flammable gasoline-based weapon, napalm was loaded into fuel tanks and was so powerful that it could create its own wind currents, triggering a firestorm that inflicted massive damage to matériel and horrific effects on human flesh.[54] Glenn understood napalm's destructive power; nevertheless, he said, "You don't think about that at the time, of course. You inhabit a small universe of you, your unit, and the enemy, and you're fighting for your own survival and that of the men you're with. It is 'destroy or be destroyed.'"[55]

On Majuro, the men again lived in Quonset huts that contained rows of canvas cots. Empty crates served as side tables. For what normally would be bathroom needs, they had to go outside, where primitive equipment did the work of a toilet. Taking a shower involved using a device to lift and dump bags of water. First, users poured enough water to wet themselves, then they put soap all over their bodies and again dumped water on themselves to rinse. Over time, Glenn realized the best showers could be taken during the almost daily drenching rain. On days when it didn't rain, he wore more than a splash of Old Spice.[56] Except in a Bob Hope USO show, the men didn't see a female during a four- to five-month period, but Glenn tried to write Annie every day from his all-male pocket of the world.[57] He estimated that he wrote her two hundred letters during his deployment.[58]

The squadron members' diet included dehydrated potatoes and eggs, canned fruit and juice, and lots of Spam, which Glenn liked. Once, weevils invaded the flour supply, so Glenn and his colleagues ate bread with baked-in bugs for a while.[59] On some days when he had no mission, he worked to improve the hut. One afternoon, he and others built an awning for their rustic home, and on another day, they played a game in which they gifted "tokens of our esteem" to bunkmates.[60]

During his time in the Pacific, Glenn had direct contact with natives only once, in a village on Tarawa. While there, the pilots bought trinkets from indigenous islanders.[61] He sometimes saw natives working with Seabees to transform the islands into suitable flight stations, but he never interacted with them.

In November 1944 the squadron relocated briefly to Roi-Namur and later that month to Kwajalein, which was the largest atoll in the Marshall Islands and had previously been the site of fierce fighting. Americans had eventually captured Kwajalein, but the battle had stolen the island's natural beauty. There were few trees and little foliage when Glenn arrived. This island had no Quonset huts, so the pilots slept in tents along the lagoon's edge, and at high tide, water slid under the tent flaps.[62]

Glenn had reached the Central Pacific after many of the big battles had been fought, but his time there was not quiet. During his last full month of service in the Pacific, January 1945, the Fourth Marine Aircraft Wing, which included Glenn's squadron and two more,[63] "flew 322 sorties, dropped 333,160 pounds of bombs, strafed with 13,550 rounds of .50 caliber, 3,500 rounds of 22mm, and 48 rounds of 75 mm shells."[64]

When the end of his service in the Pacific came in February, Glenn had flown fifty-nine successful missions.[65] A February 4 citation recommendation stated, "He is the type of pilot who feels left out if his name is not on the flight schedule, whether the mission is a strike or a routine patrol."[66] As Glenn headed home, he expected to come back to the Pacific for an eventual invasion of Japan. After working his way through the red tape of returning from military service abroad, the returning warrior was touched by the "looks and words of gratitude" that his uniform attracted.[67] He delighted in his reunion with Annie and his family in Ohio.

Then Glenn reported to Cherry Point in North Carolina but soon took a temporary assignment testing new planes at Patuxent River, known as Pax River, in Maryland. One night, his plane was being repaired, so he was napping in the ready room when a call came through from the operations office. The person calling wanted to know whether Glenn was there. "Yeah, this is Glenn," he said. The voice on the other end improbably asked, "Are you sure?" Glenn confirmed that he was absolutely certain. He rushed to the operations office and learned that a plane had crashed and the pilot, known to Glenn as "Bubba," had mistakenly donned Glenn's parachute instead of his own. The operations office had been preparing to notify Annie of Glenn's death.[68] Despite this scare, the Glenns enjoyed their time in Pax River, where they were in May when the war ended in Europe. A few months later, they returned to Cherry Point and rejoiced when Japan surrendered in August.

John Glenn was home, and the war had ended without him. He did not fight in the biggest battles or score the greatest victories, but like all war veterans, he had returned a changed man. He embraced a clearer sense of purpose, enhanced confidence in what he could achieve, and hope for a better day. The small-town boy was not lost, but he now became a smaller part of a deeper, more thoughtful man who would never stop growing.

Looking back decades later, Glenn saw his first war as totally different from the subsequent Korean and Vietnam Wars. "World War II was really the last war where we were a united country, completely clear with what we were trying to do," Glenn said in 1997. "There was no doubt in anyone's mind that the forces we were up against in World

War II, whether in Europe or in the Far East, wanted to take over the world."[69]

Fifty years after the attack on Pearl Harbor, Senator John Glenn wrote a poem about the attack and the sacrifices made by Americans during the war. Entitled "We Will," it ended with these lines:

> Our nation's future lies ahead
> A tribute to our honored dead.
> We pledge to them we'll keep it free,
> This land of hope, of liberty
> God rest their souls, through His great might.
> God grant us guidance to do the right.
> We will keep faith—
> —we will.[70]

These words reflected the powerful impact of World War II throughout Glenn's life.

Returning home, Glenn rejected his father's offer to join the family plumbing business and resisted attempts by Annie's dad, Homer Castor, to lure him into dental practice. He wanted to fly, especially if he could continue as a marine pilot. Because he came through navy cadet flight training, his commission was in the Marine Reserves, but before the final surrender, he had applied for a regular commission in the Marine Corps. The commission came through in the autumn of 1945. Soon afterward, he and Annie welcomed the first addition to their family. Annie went into labor in the middle of a stormy night. In a torrential downpour, they rushed from Cherry Point to the nearest military hospital at Camp Lejeune. John David, called Dave, was born at 6:30 AM on December 13.

Three months later, Glenn was transferred to El Toro near Irvine, California. He joined the squadron known as the Death Rattlers and became its operations officer. The war's end had led the Marine Corps to discharge thousands of marines, which meant that pilots often found themselves without the aid of enough capable mechanics and maintenance workers. To fly, they had to repair and maintain their own planes. Glenn had experience working on the cars his father sold, but the two-thousand-horsepower Pratt & Whitney R-2800 Double Wasp radial engine was a much more critical test of his mechanical skills. Understandably, Glenn felt less secure in a plane he had repaired himself than in an aircraft readied by qualified mechanics. In the postwar years, his unit became famous for flying Corsairs at air shows around California, while crowds watched them with wide eyes and open mouths. Glenn liked the performances for practical reasons: they forced the pilots to fly in formation, a skill that could deteriorate during peacetime.

Being in California enabled John and Annie to enjoy the outdoor life. They went horseback riding, hiking, and sometimes dove for abalone, which they cooked at home. Though Glenn was happy with his assignment, he knew he soon would be deployed overseas again.

As he often would throughout his lifetime, Glenn took the initiative, volunteering for special duty in China, where General George C. Marshall was trying to negotiate peace between Chiang Kai-shek's nationalist forces and Mao Zedong's burgeoning Communist movement. By this time, the Communists controlled most of the area north and west of what we now call Beijing. (At that time, it was anglicized as Peiping.) Both army and marine troops were there to provide needed protection for Marshall and his colleagues. "It was supposed to be a short tour, something like three to six months max," Glenn recalled. He hoped that prediction would be accurate. He was eager to return home; another baby was on the way. (He later said, "That turned out to be a mistake because about two years later, I got back."[71]) Army officers in China had their families with them, but marines did not have that luxury.

Upon arrival, Glenn was shocked by deplorable poverty and amazed by the obvious corruption that contributed to the burdens of the poor. A United Nations agency dropped bags of food in rural areas where Chinese were starving, and those bags often showed up on the streets of Beijing for sale within days. Glenn felt pity for rickshaw drivers who owned only their vehicles and survived on a diet of rice. Hungry Chinese walked six miles to the military base to scavenge mess hall garbage for food.[72] Glenn admitted decades later that he was surprised the Communists had not triumphed earlier, given the atrocious living conditions.[73]

His squadron flew daily patrols over northern China. Using a grid pattern, pilots were under orders to locate evidence of Communist sabotage, such as damaged power lines, bridges, and railroad tracks. When not on duty, Glenn and his fellow pilots explored the city with caution. The Communists threatened to kill one marine a day while they stayed in China. At first, marine commanders didn't take the risk seriously, but then "there were two days in a row where marines were killed," Glenn recalled.[74] After that, marines were restricted to base, and later, they could leave only when traveling in pairs. Once, when some of the troops traveled outside Beijing for gunnery practice, Glenn got his first chance to drive a tank and proclaimed that it was "a lot of fun."[75]

During the winter of 1946–47, the peace teams abandoned talks. Flying high above, pilots followed their withdrawal by rail until they reached the port of Qinhuangdao, where the Great Wall dead-ends at the Pacific. From there, the negotiators were transported out of the country. Glenn helped to fly military equipment from China to his squadron's new base in Guam. During that mission, they had a two-week stay in Shanghai. The navy's Criminal Investigations Division was working there to save Americans' lives by keeping them away from dangerous areas, such as the docks. Nightly, investigators went into opium dens to make sure no American sailors were there. The ever-curious Glenn asked whether he could accompany them one night. He remembered walking downstairs into a basement. "It was like something out of a cheap Humphrey Bogart movie, literally, with the smoke level down there. . . . What they were looking for was just to see whether there were any sailors there, but when they were down there, they covered each other. They didn't leave their backs exposed much." The officers carried submachine guns to protect themselves and each other.[76]

When they left Shanghai on the way to Guam, bad weather forced the pilots to stay over in Okinawa. While he was there, Glenn learned that his daughter, Carolyn Ann (called Lyn), had been born. He was delighted. Then news arrived that Annie had been hospitalized with an infection. His family urged him to take emergency leave. New Concord was five days away, and by the time he arrived there, Annie was home and on the road to recovery. After a short respite with his family, Glenn left for Guam.[77]

As operations officer, Glenn established training schedules to prepare the team for flight inspections. Essentially, "we were on Guam for whatever might break out, out there in that part of the world," Glenn remembered.[78] Fears of Communist subversion in Asia were growing with Mao and his forces grappling to control the giant, China. Fortunately, Guam's beautiful weather made it an ideal place to train.

Navy personnel on the island lived in Quonset huts built by the Seabees, and officers lived with their families, but no such facilities existed for marines. After many marine complaints, the navy captain who was the base's commanding officer agreed that although the Seabees no longer were available, marines could use old Quonset kits to erect their own huts. After daily flights, marines worked to construct their own future homes. Believing that his tour would end soon, Glenn did not immediately build a hut for his family, but he helped others. After the marines had completed twenty-five to thirty huts, they began filing requests for their families to join them. Then the base commander announced that the navy was going to requisition half of the newly built huts. "That came close to creating a riot," Glenn recalled.[79]

Soon, a marine general became the base's commanding officer, and he responded to protests, allowing the leathernecks to keep the huts they had built. About five months later, with no word about returning to the mainland, Glenn sought permission for his family to join him. Annie traveled across the United States and onward to Guam with her two small children. Glenn made arrangements for a used Buick Roadmaster to be shipped over as well. The family's Quonset hut was far from luxurious, but they enjoyed a chance to live together in an exotic locale for two or three months before Glenn received orders to return home. The voyage back across the Pacific was memorable; they traveled in rough seas, and one big splash came through the porthole, drenching much of their small compartment, even hitting young Dave, who was on the top bunk.[80]

After a short leave, the family headed for Corpus Christi, where Glenn looked forward to piloting again. To his surprise, he received an assignment to serve a ninety-day stint as a night shore patrol officer downtown on a 6:00 PM to 2:00 AM shift. The Corpus Christi base was huge. In 1944 it had become the largest naval air station in the world, and though the military had shrunk since the end of the war, sailors and marines still filled the city streets at night, often finding trouble. When local police discovered members of the military involved in an incident, they called Glenn's office so that he could extricate them from their predicament or arrest them. He had only two cells for jailing rowdy young men. Sometimes the troublemakers were armed, usually with knives, not guns. Shortly after Glenn took the job, he went out to a disturbance with a tall and husky navy chief, who had been on shore patrol duty for quite a while. They reached a disorderly mob, and Glenn leaped from the truck. "I was pulling people, trying to get into the center of this crowd to get the thing broken up. . . . I almost got in there, and I looked around to see where Botterman was, and it turned out he hadn't left the truck yet. . . . So I beat a strategic retreat until we got better organized."[81] He learned a lot from Botterman, although years later, Glenn remembered only his last name. On one occasion during this unusual assignment, Glenn was flabbergasted when a prostitute came into his office to file a complaint against a serviceman who had failed to pay for her illegal services.

When that assignment was over, Glenn entered a six-week program in an Instructors' Advanced Training Unit at Cabaniss Field, also in Corpus Christi. He learned standardized teaching techniques, studied weather conditions, and expanded his knowledge of aircraft maintenance. He moved to a training unit, where he was expected to teach others to fly Corsairs and Hellcats. After only a few days there, the Instructors' Advanced Training Unit asked him to return as an instructor of instructors, so he did. While following established lesson plans, he looked for ways to weave his own experiences into each class.

Always looking for chances to engage in unique experiences, Glenn sought an opportunity to take three months of air force instructor training so that he could compare the two services' programs. He underwent air force training at Arizona's Williams Air Force Base, and while there, he took advantage of his first opportunity to fly jets. "We just had the first few jets coming into the Navy and Marine Corps back at that time, just getting started with the very first squadrons, and we did not have jet training in the training command as such yet," he said.[82] "In the [Lockheed] P-80 [Shooting Star] out there we were traveling, just on average, probably twice as fast as we traveled in the prop planes, prop fighters, back at Corpus Christi. . . . It was just a whole different kind of flying."[83]

Beyond getting his first taste of jet piloting, what he learned during this assignment was that air force trainers concentrated on testing instructors as pilots and applied much less attention to weapons procedures such as bombing practice or gunnery drills. He wrote a report to the chief of naval air advanced training, saying, "The problem of indoctrinating the new instructor is approached from two entirely different angles by the Navy and the Air Force. In the Navy program, we assume that the newly arrived instructor is a competent pilot and that our mission is not to teach him to fly, but is to teach him to instruct. . . . The Air Force approaches the instructor check-out not with a view toward teaching him how to instruct, but with a view toward checking on his flying ability. It is a flight proficiency program."[84]

During the time they were based in Corpus Christi, the Glenns bought their first house in early 1949 for about $12,500. It had three bedrooms but no air conditioning. They loved its fireplace, probably because it reminded them of their midwestern roots. Annie again found work playing the organ at a local church. They liked the area but were not enthusiastic about the scorpions that crept into garages or the tarantula John found on their doorstep.[85]

Glenn's next challenge was the Naval Aviation School of All-Weather Flight, a three-month course. "The school was known throughout everywhere," he said.[86] During training, the pilots wore goggles that eliminated all visual clues about their position. He received advanced training on flying by instruments alone, and again, at the course's end, he was invited to remain as an instructor. Flying small planes in sometimes problematic weather, the pilots typically came in for landings by being positioned in a "stack," and each plane would be cleared for landing individually. On one cold and stormy day, the planes began icing up, and suddenly, they lost contact with the radio facility that usually picked up each plane's navigational fix and guided it to a safe landing. "We were up there with people in the stack and no way to get down," he recalled. "People literally fanned out all over south Texas. I think we had airplanes wound up in about five or six different fields." Each pilot stayed at his assigned altitude to avoid collision.[87] That aerial scramble represented an extraordinary event in his flying career.

At the end of his time in Corpus Christi, Glenn received an assignment he dreaded. In July 1951 he found himself at Amphibious Warfare School at Quantico, Virginia. The last thing he wanted was to give up flying and be stuck on the ground. Nonetheless, the Marine Corps required officers to take the six-month course. While participating, he commanded a ground division during war games. He also learned about field artillery, a key component of marine assaults on the battlefield, and he got his first training on deployment of tanks.

When Amphibious Warfare School was over, Glenn hoped for an assignment in the air; instead, in December 1951 he was placed on the general staff of Quantico's commanding officer. He was in the operations section, where he was responsible for scheduling use of live firing areas in the base's "Guadalcanal area" and guaranteeing that vehicles did not enter the area where weapons were being fired. He hated it.

Glenn wanted to fly, but he admitted decades later that he tried to embrace every assignment he was given. Always ambitious, he had his eyes on a big job. "Anybody in the Marine Corps, any marine officer, hopes that someday he'll wind up as Commandant," he mused.[88] Therefore, he wanted to maintain a good record, even when he was marooned on the ground.

After giving up their first house in Texas, the Glenns were disappointed to find that Quantico had no base housing available. They put their names on a waiting list and temporarily settled in a run-down house. Then they moved into half of a Quonset hut, while another family resided in the other half. With only fiberboard to divide the two families, "you could literally hear each other breathe," Glenn said.[89] The huts in an area called Midway Village had no central heating, so a potbelly stove was the only heat source. The Glenns thought their half hut would be a short-term home until base housing became available, but they spent six months there, including the entire winter.

As Glenn tried to become accustomed to ground duty, events on the other side of the world made him even more anxious to return to flying. On June 25, 1950, about seventy-five thousand North Korean soldiers had invaded South Korea, crossing the thirty-eighth parallel boundary separating the two nations. Communist North Korea sought to overrun its nominally democratic, anticommunist southern neighbor, which was allied with the United States and other capitalist nations. The United States joined fifteen other United Nations members in defending South Korea. Glenn watched as his nation marched into war just a month after the shocking flood of North Korean troops into the south.[90]

When General Douglas MacArthur led UN troops into North Korea, Mao sent Chinese troops into the conflict, increasing the threat of a ballooning war. Chinese troops forced allied troops back into South Korea. To regain the upper hand, MacArthur wanted to bomb China and draw Taiwanese troops into the conflict. President Harry Truman rejected both ideas as MacArthur lobbied others to support his position. Fearful that MacArthur would recklessly escalate the confrontation into a world war and outraged by MacArthur's insubordination, Truman fired MacArthur, although many Americans still viewed the triumphant World War II general as a hero. After this dramatic confrontation, the war increasingly became a series of one-step-forward-two-steps-backward battles for a few hundred yards of ground. As casualties continued to rise, Western leaders argued that they had to protect South Korea to block the Soviet Union and Communist China from overrunning Asia.[91]

While serving on the Quantico commander's staff, Glenn applied each month for a war assignment, and the answer was always the same:

request denied. "After several episodes of this, I was called into my boss's office one day, who was a marine colonel, a ground officer, and he told me in no uncertain terms that I was not to put any more letters in, that when the Marine Corps decided I would go to Korea, I would go," Glenn said.[92] At home, at least, there was good news: the family eventually found a home in base housing.

Then Glenn received notice that he was being reassigned to South Korea. First, he had to go to Cherry Point on temporary duty to receive jet training that complemented the experiences he had enjoyed while flying with the air force. After that, he took leave in Ohio before traveling to the West Coast for departure to South Korea. The family relocated to New Concord, where Dave was in first grade.

John Glenn was not a man with a lust for war, but he had learned that he had a talent for flying in combat, and he looked forward to sharpening those skills in a new war against a new enemy. The addition of jets to the American arsenal offered a challenging opportunity to expand his flight expertise. He was ready to go to war.

3

OL' MAGNET TAIL

In the peace of an Ohio day, John Glenn flew above farmland south of Columbus and scattered the ashes of a friend and fellow pilot who had died in the war he was about to join. Performing this favor for his friend's widow was a solemn duty and a tangible reminder of war's capricious and deadly potential.[1] And yet, Glenn "looked forward to going to Korea," he later said.[2] And after he returned to his life as a wartime fighter pilot, his zeal for the job led others to characterize him as reckless, while some thought that he merely felt a duty to protect others despite the risk to himself.

One of the happy benefits of his new assignment was a frequent opportunity to fly with baseball star Ted Williams as his wingman. The Boston Red Sox great was in the inactive reserves when the marines called him to fight in Korea, as he had in World War II. Some fans protested this second interruption to his baseball career. In 1942 and 1947, Williams had won the American League Triple Crown, topping all competitors in batting average, runs batted in, and home runs. In 1949 he managed to get on base in eighty-four consecutive games.[3]

Williams himself "always felt he'd been called back in part because of his fame," and he felt "picked on."[4] The baseball titan, who arrived in Korea twelve days before Glenn, became a close friend. Among the things they had in common, fellow pilot Marsh Austin said, was that "neither Ted nor John Glenn ever really drank or smoked cigarettes at all."[5] Williams had two nicknames within the squadron—Big League, a reflection of his true status, and Bush, a joke identifying him as bush-league.

In this war, pilots typically flew in support of ground forces. With the two sides often fighting in close proximity, a big part of the job was making sure that no US weapons—rockets, bombs, bullets, or napalm—hit allied troops. Without sophisticated guidance systems, flight leaders learned to recognize the targeted terrain based on charts and aerial photos.[6] Sometimes fliers encountered weather they characterized as "milk-bottle-like conditions": light fog sometimes made it impossible to see other members of the squadron or anything on the ground. Often, American jets attacked in an "around-the-clock approach," forming a circular motion with a line of fighters. Each plane in the circle struck in sequence and the formation rotated, thus making it more difficult for antiaircraft weapons to anticipate each plane's angle of attack.[7] When not committed to ground support, pilots also soared above North Korea in search of targets whose destruction would hamper the enemy's ability to wage war in South Korea. Supply transports, bridges, tunnels, and railroads became prime objectives.

Glenn and Williams were stationed at K-3 at P'ohang in South Korea's southeastern coastal area, where members of the squadron lived in Quonset huts far from active battlefields. Glenn saw North Korea as a surprisingly efficient enemy. "They were very, very good at putting huge crews in at night and repairing roads and bridges and railroad bridges . . . and having them operational again" the next day or within a few days.[8] "We had no way of attacking them at night," he added. To make up for that handicap, US forces would go out early in the morning and "try to catch them still out on the road."[9]

On one early morning flight, Glenn and Williams came dangerously close to a catastrophic error. On reconnaissance flights, one pilot flew low, while the other rose higher and behind the first so that he had a fuller view of what lay ahead. The two pilots swapped positions about every ten minutes. On this day, they flew their entire course without seeing anything to attack in North Korea. "It was very disappointing to be all keyed up for a big mission, fly the mission and make no contact," Glenn later said.[10] With no target in sight, pilots were supposed to fire any remaining high-velocity aircraft rockets (HVARs) before returning to base. HVARs were unguided weapons, and their proximity charges made landings dangerous. In the absence of a perfect target, Glenn crossed back over into North Korea and pummeled an abutment. Williams imitated his actions, but he forgot to turn a switch in the cockpit. Consequently, his rockets didn't fire. Glenn and Williams then headed back into South Korea, where Williams fired at the ground.

Glenn was almost apoplectic; the chart in his cockpit indicated that Williams had fired his rockets toward territory occupied by allied troops. As the flight leader, Glenn would have faced an automatic court-martial if Williams had attacked "friendlies." When they returned to base, Glenn rushed to the operations shack and positioned himself in front of a large map that showed what areas the North Koreans controlled. To his enormous relief, the line had moved five hundred to seven hundred yards, and the spot Williams had hit was currently occupied by the enemy. For the rest of their lives, Glenn teased Williams about almost ruining his career.[11]

As he became more accustomed to being at war again, Glenn's zealous attacks raised some concerns among fellow pilots. Once, he spotted an antiaircraft gun on a run, so after he hit his target, he ignored protocol and returned to destroy that weapon. Unfortunately, as he was pulling up from a deep dive, he was struck by another antiaircraft weapon. It hammered his plane's tail, causing a malfunction. He thought he was going to crash into a rice paddy, but with a struggle, he maneuvered back to base. When he did, he was shocked to see the damage; the hole was big enough for him to put his head and shoulders through it.[12] Pursuing a second target after hitting the first was a violation of accepted procedures. He just couldn't resist. He wrote an eighteen-stanza poem about his adventure. It ended this way:

If you would bold and older be,
Be "Tigers" all, tis true,
But on a bombing run just make,
One run, just one, not two. . .[13]

Decades later, an interviewer asked Glenn why Williams had referred to him by saying, "The man is crazy." Glenn said Williams was commenting on his actions in combat. "I was rather intent on war making when I went out there. . . . I won't say I was trying to win the war all by myself, but I took it all very, very seriously, and things such as a second run on an antiaircraft position, which wasn't exactly—well, I won't say it was foolhardy, but you put yourself in a lot of danger and you knew it."[14]

Enemy fire hit Glenn five times in World War II and seven times in Korea. Williams later recalled, "We called Glenn 'Ol' Magnet Tail' because his plane was hit so many times in combat."[15] Glenn returned from one mission with 250 holes perforating his jet's surface,[16] and he left that battle somewhat nonchalantly, saying, "I'm going to ease out of here."[17]

A more critical account in *U.S. Marine Corps Aviation Since 1912* says Glenn sometimes recklessly "put his life in grave danger and [this recklessness] earned him the early sobriquet 'Magnet Ass,' because of his seeming ability to attract flak." Too often, the book concludes, Glenn limped back to base.[18] However, he was proud of his record: "I never had an airplane quit," he proclaimed. "I never had to bail out."[19] At the same time, Glenn admitted that he often dived lower than necessary. Besides the possibility of crashing or being hit by antiaircraft fire, diving pilots faced the threat of being destroyed by the bombs they had just dropped. To avoid that possibility, they had to use delay fusing to protect themselves.[20] Glenn's likely reasoning for embracing danger was well described by one of his World War II commanders at El Centro, who said, "Marine training makes every marine more afraid of letting his buddies down than he is of getting hurt himself."[21]

Glenn was not a cowboy simply looking for a wild ride. He believed in the Korean War and thought the United States was right to be there. In a letter home to his daughter, he explained why he was so far from home: "In our country, Lyn, we all believe that people should be allowed to have their own homes, to work where they want to work, and do the things they want to do. But some countries have people in their governments who do not act that way. They want always to be the bosses." He was fighting, he told her, to protect the rights of all people in non-Communist nations—and that's why he had missed her sixth birthday party.[22]

Glenn's comrades in the squadron liked his down-to-earth attitude, journalist Eloise Engle wrote. She recalled that when he was on liberty, he enjoyed swimming, reading, and taking long walks.[23] Before taking leave in Japan, Glenn collected money and brought back a piano for the officers' club.[24]

Perhaps in a foreshadowing of his political career, he also found time to write opinion pieces for military publications. One article, entitled "Information, Please," argued for establishment of a Tactical Air Board to keep all pilots up to date on the latest tricks of the trade.[25] In another, "Before the Run," he responded to reports of a plan to train private pilots to be reservists through two summer vacation periods between college years rather than giving them strict military training. Glenn opposed these "summer cruises."[26] He also sent the *Marine Corps Gazette* a satirical fake interview with a general.[27] The *Gazette* replied, saying, "We got a good chuckle from it but naturally will be unable to use it in the *Gazette*, which I am sure you suspected when you sent it in."[28] In retrospect, the truly ironic part of this article was the fictional general's discussion of a "shuttle principle" that would enable a plane to fly around the world, drop a bomb, and continue making a complete orbit. Though Glenn never dropped a bomb from space, his orbital spaceflights bore more than a little correlation to this imaginary scenario.

On one big Korean operation, Glenn's superior officers asked him to provide the choreography for a massive napalm strike by both squadrons at his base. US forces wanted to destroy a large area in North Korea filled with both people and supplies. Glenn mapped out an assault that assigned planes to approach the area from different directions, confusing the antiaircraft gunners. All flew low and dropped their bombs at the end of the drop zone so that rising smoke would not blind those flying behind them.[29]

Toward the end of his duty in Korea, Glenn volunteered for a special assignment: He asked to fill an opening in an air force fighter-interceptor squadron. Each air force squadron left one position open for a marine, and Glenn desperately wanted an opportunity to fill one of these slots. He was eager to join a squadron flying air force F-86s against Soviet-made MiGs. Marine pilots' ground-support role made the possibility of one-on-one combat highly unlikely. He hungered for the challenge of old-style dogfights at incredibly fast speeds. Before he went to Korea, he had asked air force colonel Leon Gray, whom he had known at Williams Air Force Base, whether he could visit Otis Air Force Base in Massachusetts to prove himself on the F-86, just in case he got a chance to fly one in Korea.[30] The politician buried not so deep within Glenn knew how to pull strings, and that's what he did.

The air force's goal was to keep Soviet-made MiGs based in China from interfering with allied operations. Often on major air-to-ground operations, the F-86s established a four-plane screen to block MiG penetration of the field of battle. The planes were supposed to hold their relative positions in an expanded formation that placed them several hundred yards apart, but when one US plane "bounced"—started an attack run—maintaining the screen became difficult. "If you had an enemy aircraft headed towards you going 550 miles an hour and you're headed toward him at the same rate, your closing rate is 1,100 miles an hour, which is very, very fast. You have split seconds to make a decision," Glenn said.[31] The speeds were great, and the technology was improving. They had a computing sight to help with targeting the enemy. Nevertheless, he said much of the shooting still was done "by seat of the pants, sort of, or intuition."[32] There was radar equipment in the nose of the plane, but it was somewhat primitive. Ground radar, too, helped to identify friend or foe, but its station was off North Korea's western coast on Chodo Island, which was often too far away to help.[33] In an awkward mix of high tech and low tech, pilots of interceptors carried binoculars to enable them to identify fast-moving MiGs.

John Glenn in North Korea during a period when he flew in an air force squadron, and his plane received a new nickname, MiG MAD MARINE. *US Air Force*

Glenn was eager to test himself against MiGs and complained early in his service with the air force when he heard that his time in the air might be limited; news reports suggested that peace talks, which had begun in July 1951 under President Truman, soon might bear fruit before he had an opportunity to challenge many MiGs. The next morning, he found that his plane had a new name painted on it: MiG MAD MARINE.[34] Glenn need not have worried; he would make three "kills" in the closing days of the war.

One day, Glenn followed his commander, John Giraudo, as wingman in an assault on a long line of trucks in North Korea. Giraudo's plane was hit, and he lost control. He tried to get the F-86 over the ocean but failed and had to abandon his jet, parachuting into North Korea. Glenn saw him reach the ground safely and requested an air-sea helicopter rescue unit from Chodo Island. Glenn and two other pilots stayed at the spot, circling as long as they could, but rescuers never came. Eventually, the circling planes ran low of fuel and flew back to base. The last to leave, Glenn flamed out and glided to his home base's runway. He had radioed ahead to have another plane ready so that he could take off immediately and return to the skies above Giraudo's last known location. Four planes joined him, and again, the group circled, but no rescue helicopter came. Glenn was distraught, and after he returned to base, another pilot recommended that he see the chaplain, Father Dan, who had been praying with rosary beads for Glenn. "They said he was running the rosary like a monkey climbing a flagpole," he recalled.

Later, when there was a prisoner exchange, Glenn asked Tom Miller, a fellow marine officer and World War II friend who was in charge of the transaction, to let him attend. When Giraudo emerged and saw Glenn, he said, "Oh, you son of a gun, I grew to hate you." Glenn said, "Why? What happened?" Giraudo revealed that all of Glenn's circling had attracted attention, so North Koreans came looking for him. He was working his way through wooded terrain when he heard people nearby. Then his life raft, which he had hidden, suddenly inflated, and the noise attracted the North Koreans to him. His captors stuck a machine gun in his face. They never shot him, but they roughly threw him into a ditch and sat on him every time Glenn flew overhead. Giraudo "wished I would go away," Glenn later said.[35]

He scored his first "kill" leading a four-man team. He saw two MiGs flying low, so the team swooped down to attack them. The chase began about fifty miles south of the Yalu River, which divides China from North Korea. US planes were not supposed to cross into Chinese airspace unless an engagement began over North Korea and crossed the Yalu. Glenn and his wingman followed their targets beyond the river, and each hit his objective, with Glenn's MiG crashing and exploding on a Chinese runway. Flying low, Glenn came frighteningly close to that runway himself.[36]

On another day, he decided to attack four MiGs crossing the Yalu into North Korean airspace. When the planes turned, he realized there were sixteen, not four. Glenn targeted a jet and believed he had the advantage. Then the MiG went into a climbing turn. Glenn tried to match his foe's pattern, but "he still out-turned me enough that in about three full climbing turns, it went from me shooting at him to him shooting at me." Glenn's wingman managed to down the jet, but the wingman developed engine trouble. Glenn left the melee to make sure his comrade made it to safety. Along the way, six more MiGs engaged him. Almost immediately after Glenn fired on them, they turned to flee. He managed to hit the last jet, which crashed. His final kill came on a day when Glenn started an attack after seeing four MiGs. Firing heavily, he came so close to one that he almost rammed it before the MiG went into a dive and crashed.[37]

In a report on his experiences with the air force, Glenn hastened to make clear that shooting down MiGs was not a daily event. "It should be understood that many missions were flown during this period in which no contacts were made with enemy aircraft," he reported.[38] Glenn believed his combat experiences helped to convince NASA a few years later that he was fit to be an astronaut.[39] In all, Glenn flew sixty-three combat missions for the marines in Korea and twenty-seven for the air force, for a total of 149 across World War II and Korea.

In July 1953 an armistice ended the war after the deaths of roughly five million civilians and soldiers, including almost forty thousand Americans. The armistice, coming during the presidency of Dwight D. Eisenhower, allowed Korean prisoners of war to live in whichever nation they preferred and granted South Korea fifteen hundred square miles of additional territory, while instituting a two-mile-wide demilitarized zone between the two Koreas.[40]

When Glenn looked back at the Korean War years later, he called it a "forgotten war." He believed "it came too close after World War II where everybody had been so concentrated on a huge war all over the world, and this war was looked at as sort of a small regional conflict that came up within five years of the end of World War II."[41] The Korean War, like all the smaller wars that would follow, would be compared to the global conflict. The US triumph in a massive multifront world war was a tough act to follow.

In the wake of the Korean War, Glenn remained on the peninsula to finish his deployment. At that time, the military conducted a lottery that offered servicemen rest-and-relaxation leave in Hong Kong, and on one occasion in the autumn of 1953, Glenn was a lucky winner. These trips were not just vacations; aircraft carriers in the Pacific made regular port-of-call visits to Hong Kong, and the carriers' crews made room for a few land-based servicemen to ride along. The trip took four or five days, and the soldiers spent about a week in Hong Kong before returning to Korea aboard the carrier.[42]

"That was the old Hong Kong," Glenn recalled. "That was when you sat out on the front porch of the Peninsula Hotel, like you used to see in the old-time ads or movies, before Hong Kong grew as tall as it is now."[43] While Glenn and his colleagues enjoyed the glamour of Hong Kong, they did not get to sample all of the city's luxuries: every night, they returned to the carrier, where they slept in crew quarters rather than luxurious hotels.

Glenn finished his service in Korea in December 1953 and headed home. He requested Test Pilot School and was exuberant about winning the assignment, although Annie feared the risks he would take. "We had a whole new stable of aircraft that we knew about that were being designed and were coming out at that time," he said. "They included the first of the Navy and Marine Corps supersonic airplanes as well as a whole new series of jet attack and fighter aircraft." He attended Test Pilot School at the Naval Air Test Center at Pax River in Maryland.[44]

Glenn's six-month course began in January 1954, and he immediately found himself overwhelmed by rigorous academic courses. "It was like we used to joke about: Going to the academic side of test pilot training was like getting a drink out of a fire hydrant. You get a little in you and a lot all over you," he remembered. "It was a very, very tough academic course because a high percentage of the people that came in were either people who had graduate degrees already or who were graduates of the Naval Academy." Those people had studied advanced calculus, but no trace of calculus appeared on Glenn's abbreviated college transcript. A background in calculus was a requirement for admission; however, commanders had waived that requirement as a result of Glenn's combat experience. "That's one of the hardest academic things I ever did, was do calculus self-taught," Glenn recalled. "I stayed awake till two or three in the morning many, many nights, going through and learning calculus with all the little primers and background."[45]

He won his battle with math, passing flight training and moving on to an assignment with a Pax River group known as Armament Test. At around the same time, Annie and the kids arrived from New Concord at the end of the school year and settled in an apartment just off base.

Within Pax River's test pilot operations, there were several teams: Flight Test studied planes' aerodynamics, Service Test checked durability by flying a new plane as much as it would be flown in its lifetime, and Armament Test measured a plane's battle-readiness in simulated combat. Each Armament flier evaluated one or two planes at a time. "So you went out and you fired the guns and you shot rockets and you dropped bombs and you did all the things you normally would do in combat."[46] Glenn and other pilots spent a lot of time firing guns mounted on their planes. Another plane towed a banner that served as a target. Each pilot carried shells adorned with a different color of soft paint, making it possible to count the number of holes each pilot had made in the banner.

Through testing, the pilots discovered flaws in the planes—sometimes unsettling ones. For example, Glenn flew a trouble-plagued FJ-3 Fury, made by North American Aviation. It was similar to the planes he had flown as a marine in Korea, except for a few changes, such as the addition of a tailhook for carrier landings and replacement of machine guns with twenty-millimeter cannons. These alterations made the plane heavier and affected its performance. The cannons caused damage to the plane's nose, sometimes dislodging rivets and blocking one of the jet engine's inlet ducts. As a result of Glenn's experiences, engineers had to redesign the plane to strengthen its support structure. Later, "they had a few squadrons of those," he said, "but it never was a plane that was bought in any big major quantity, but for a little while it was about as good as we had."[47]

Glenn devoted a lot of time to testing a Vought fighter, the F7U Cutlass, which, in his words, "looked like a praying mantis." The craft was engineered to function on aircraft carriers. He explained, "To get an angle of attack that would come aboard ship, where it could land even on a runway, you had to have a very high nose gear." The plane was a star performer at low altitudes, but higher up, it tended to stall and spin, creating hair-raising moments for pilots. Glenn discovered that twenty-millimeter cannon fire obstructed the inlet ducts on both sides. Firing the weapons disrupted the flow of air going into the engines. "I think if you're above about 25,000 feet, when you would fire the 20-millimeter cannons, you'd put the fire out in the engine. So there you are flying a glider up there at 25,000 feet," he recalled.[48] After the problem had been diagnosed, an armaments expert was able to create a device to disperse air from the weapon so that it would not affect the engine. That worked until the third or fourth trial flight, when the device burned off, falling into one engine and setting off every emergency light in Glenn's cockpit. Pieces tore up the engine, and shrapnel perforated the fuselage, including the area around the fuel tanks. Glenn was lucky that the fuel tanks did not explode.[49] More than 25 percent of these planes were lost in accidents between 1948, when the first prototype flew, and the late 1950s, when F7Us were retired.[50]

Glenn enjoyed his work as a test pilot, though he knew it couldn't last. At the end of almost two years, the normal period of service for a test pilot, he was unhappy to be reassigned to the navy's Bureau of Aeronautics in Washington, in November 1956. Rather than disrupt the children's schooling again, the Glenns decided to continue living in the same apartment near Pax River, and John made a 105-minute commute to and from work each day. The adjustment from flying to riding a desk was burdensome, but in a way, this job grew directly out of his test pilot experience. At Pax River, "you flew the airplane; you made the recommendations [for changes] on the basis of what would make the airplane better without regard to what the cost would be," Glenn said. Considering the cost was now his job in the Bureau of Aeronautics, "so that became something different."[51]

Glenn served in the fighter design branch and spent huge chunks of time in meetings with people working in other branches of the Bureau of Aeronautics. He frequently traveled to aircraft-manufacturing plants around the country. He was able to focus primarily on planes he had flown, but he said, "It was a different kind of work. It was all paperwork. So I didn't enjoy it nearly as much as I had my previous assignment." Despite displeasure with his deskbound existence, he recognized the significance of what he was doing: "This time was very important because without the approval from the Bureau of Aeronautics and the assignment of money to the contractors to fix whatever the problem was—that's just what you had to do to make a fleet-ready airplane or a Marine-ready airplane."[52]

The demanding Glenn took his work seriously and sometimes tangled with his future NASA boss Chris Kraft, who was then a project engineer assigned to unravel ongoing problems with the navy's new supersonic F8U Crusaders. Once Kraft bluntly told Glenn that the F8U "is dangerous." Glenn responded that he had flown that plane without experiencing the problems Kraft was citing, and in Kraft's memory, Glenn told him, "You don't know what you're talking about." Glenn had a gut instinct to accept what the manufacturer was saying, which was not always the best course to follow. On this particular day, Glenn knew that a pair of F8U pilots were in the building, so he asked them to join the conversation. When the group gathered, both pilots acknowledged the problems Kraft was describing—and Glenn quickly accepted their assessments. Urgent repairs had to be made on all F8Us in service, but Vought, the manufacturer, developed a quick fix that resolved the issue. Later, the discovery of more problems with the F8U led to grounding the entire fleet until the malfunctions could be corrected.[53]

Overall, John Glenn was restless, hungry for a challenge and homesick for the skies. Then an idea emerged that would give him a challenging flight. His goal was to set a record for a transcontinental trek from California to New York. Glenn knew that it was time for engine tests on the new F8U Crusader; therefore, he proposed that the plane be tested in one big flight rather than in a lot of small ones.

To win approval, Glenn took the idea to the chief of the Bureau of Aeronautics' Admiral Robert E. Dixon. The plan might have died there, but Dixon was intrigued and said that he wanted to consult with others. The experiment was named Project Bullet, a title that Glenn called "eye-catching or ear-catching."[54] Dixon's staff raised questions, but as it became viewed as a valid high-power engine test, Glenn's proposal won approval.

Details of the flight were challenging. The navy and Marine Corps had no high-altitude refueling jet tankers. The air force had them, but it also owned the existing record for a transcontinental flight. Ultimately, the air force refused to share any planes. As a result, Glenn had to use old turbo prop planes as tankers. Such planes could rise to only twenty-five thousand feet. To make this work, Glenn had to cruise at about fifty thousand feet and then slow down and descend to hook up with a tanker for refueling.[55]

Glenn planned to run wide open with the afterburner, so he would be consuming fuel like a famished teenager devouring a bag of chips, and he would have to decrease speed multiple times to get fuel. The fuel transfer would not be easy, so he had to practice repeatedly before he could make the flight. On one occasion, when he tried hooking up to a tanker north of Dallas, there was "a big blast of black smoke" from the tanker. He immediately pulled out but stayed close to see what the problem was. Both engines of the tanker were losing power, and when the plane dropped to three thousand feet, crew members began to bail out. The pilot aimed the tanker at an unpopulated area before he became the last of three crewmen to abandon ship. A later investigation showed that someone had mistakenly filled the plane's tanks with jet fuel instead of aviation gasoline.[56]

After many practice runs, Project Bullet began with a takeoff from Naval Air Station Alameda on July 16, 1957, a day that promised good weather from coast to coast. The official clock started ticking when the Crusader flew over the federal building in Los Angeles and ended when it crossed above the federal building in New York. The organization that certifies flight records, Fédération Aéronautique Internationale in Paris, required only that the flight commence within twenty-five miles of Los Angeles' federal building. To raise the likelihood of breaking the record, Admiral Dixon decided two planes should make the flight in case one ran into trouble. Navy Lieutenant Commander Charlie Demmler roared down the runway in a second Crusader just twenty minutes after Glenn's departure. At the first refueling, Demmler hit the funnel-shaped device at the end of the fuel line and broke it, so he couldn't refuel and had to abandon his mission in Albuquerque. Glenn continued at supersonic speeds. He refueled over Kansas without difficulty, but when he descended to find the tanker over Indianapolis, he could not see the assigned plane. A backup tanker spotted him and gave him the correct bearings for his tanker. Glenn couldn't wait, so he used the backup tanker just as his fuel tanks were becoming dangerously close to empty. Coincidentally, Glenn's flight soared above the area where he had grown up, and rocked his hometown with a sonic boom. A neighbor ran to his parents' home and yelled, "Johnny dropped a bomb! Johnny dropped a bomb!"[57]

As Glenn descended toward New York, sonic booms broke some windows. His fuel supply was very low when he landed at Floyd Bennett Field in southeast Brooklyn, "and I remember I had some doubt that the thing might run out of fuel taxiing in," Glenn later said.[58] He worried that such a halt would be embarrassing with a crowd of people awaiting his arrival, but he had no trouble. He broke the speed record. At an average speed of 726 mph, he shaved about 23.5 minutes off the previous record.[59] He finished in 3 hours, 23 minutes, 8.4 seconds,[60] traveling 2,460 statute miles.[61] Despite his largely successful flight, Glenn failed to achieve two goals that he had set for himself: an average speed above the speed of sound and beyond the pace of Earth's rotation.[62] In addition to setting a speed record and testing a new plane, he had conducted a coast-to-coast photographic mission. He flew the photographic version of the Crusader, which was lighter than a standard fighter. It had mounted cameras that took photos of the United States from takeoff to landing.[63]

When he stepped out of the plane, Glenn was happy to see Admiral Dixon alongside Annie, Dave, and Lyn. He gave Dave a "supersonic knife" that he'd kept in his shin pocket and presented Lyn with a "supersonic cat" in the form of a Siamese cat charm.[64] Many reporters, photographers, and TV cameramen were on hand, and for the first of several times in his life, Glenn was surprised by the attention he had attracted. "I knew there would be some attention because I had seen when aviation records were broken before, there was always something in the paper or news magazines, usually a short spread of some kind. It seemed to take over a little bit, to get more attention, whether it was a slow news day otherwise, or what the reason was, I don't know."[65]

Glenn, who by now was a Marine Corps major, was awarded a Distinguished Flying Cross citation by President Eisenhower.[66] He also received a congratulatory telegram from his Korean War wingman Ted Williams. "Congratulations on record. I am big shot now telling everyone I flew with you in Korea,"[67] the nationally renowned hitter wrote sarcastically. In its coverage of the flight, the *New York Times* made a point of noting that, at thirty-six, Glenn was "reaching the upper age limit for piloting complicated pieces of machinery through the air at speeds that can only be compared with those of sound and of the Earth's turning on its axis."[68] Obviously, reporters had no inkling of what lay in John Glenn's future.

The Crusader's manufacturer put a spotlight on Glenn's flight by producing a special edition of the *Vought Vanguard*, its internal publication, on the day of the flight, and several days later, one of its columnists wrote, "Well into our third decade of striving in true Horatio Alger fashion to achieve fame and status as a full-fledged celebrity, we doff our pinwheel-topped beanie to Major John Glenn Jr."[69]

The next day, Glenn experienced another unexpected event, while shopping with Dave. A woman recognized him and asked whether he would be interested in appearing on *Name That Tune*, a popular television quiz show.[70] A music lover, Glenn was interested, but he expected a negative response from the military and was surprised when he got the go-ahead. He partnered with a ten-year-old boy, Eddie Hodges, who was an actor and singer. They competed on the show for several weeks and went all the way to win the top prize of $25,000 for a two-person team. A video of one of their appearances shows the pair

John Glenn with Eddie Hodges as they compete on *Name That Tune*.
Getty Images

coyly taking their time to name several songs in a "Golden Medley." In
the last seconds, they identified the final song. Glenn demonstrated a
great deal of stage presence in this early TV appearance, with no signs
of nervousness and a twinkle in his eye as he and his young partner
playfully let the clock run down to add to the suspense.[71]

During one appearance, the master of ceremonies, George DeWitt, asked Glenn to comment on the Soviet launch of *Sputnik*, the first manmade satellite, in October 1957. The future astronaut expressed open admiration for the Soviet achievement, saying, "What the Russian scientists have done is really out of this world. This great experiment of the Soviet Union is a tremendous advance for international science. This is the first time anybody has been able to get anything up that high . . . and keep it up. George, I feel sure that within Eddie's lifetime, men will be making trips to the moon."[72] (Glenn had yet to meet John F. Kennedy, a man of loftier ambitions; Neil Armstrong would stand on the moon when Hodges was a young man of twenty-two and when Glenn himself was a mere forty-eight.)

For the first but certainly not the last time in his life, Glenn received fan mail. An eight-year-old girl from Rural Retreat, Virginia, wrote to praise his new speed record and his appearance on *Name That Tune*.[73]

At the end of the 1957–58 school year, the Glenns moved to Arlington, Virginia, to reduce John's daily commute. He and his old marine buddy Tom Miller had found side-by-side lots for new houses.

When Glenn was questioned about *Sputnik* on *Name That Tune*, he had no idea how quickly the Soviet success would change priorities for his nation and for himself. The United States stood at the brink of great exploration, and Glenn would become a sunny representation of what America could achieve. His days as even a semianonymous face in the crowd were about to end. He soon would become known to almost every American.

4

RACING TOWARD SPACE

The Cold War found a bloodless battlefield in the frigid vacuum of space. Neither the United States nor the Soviet Union wanted to risk a direct military confrontation that might trigger a disastrous nuclear war; therefore, space became the Cold War's most visible and most tangible field of combat.

Shining as it moved across the night sky, *Sputnik* and its more visible spent rocket, which also orbited, seemed to mock Americans and flaunt Soviet superiority in late 1957. A second Soviet satellite entered orbit just a month later. It circled Earth for 103 days, though its doomed canine occupant, a Moscow stray named Laika, succumbed within hours.[1] When the United States entered the game, TV viewers watched in horror as the US Navy rocket Vanguard exploded spectacularly on liftoff in December 1957.[2] Early the following year, the army managed to launch the first US satellite, *Explorer 1*. The United States launched five satellites in 1958; three succeeded, and two failed.[3]

In these years, fear acquired a new dimension as Americans wondered whether a totalitarian government could offer greater productivity in science and weaponry as the Germans had in World War II rocketry and as the Soviets were demonstrating in space now. In a way, this question set the course of John Glenn's future. He had no idea that within a few short years he would become a Cold War personification of democracy—a grinning, freckled, balding man who would make Americans feel like winners again.

On October 1, 1958, the National Aeronautics and Space Administration set to work. At first, the agency briefly considered opening astronaut selection to adventurers of all types—mountain climbers, deep-sea divers, wilderness explorers—but President Eisenhower, who had planned the World War II Allies' massive 1944 D-day invasion of German-occupied France, knew something about efficiency, and he identified a better way to choose. He recommended limiting the list of candidates to military test pilots,[4] all of whom had some engineering training and experience dealing with unexpected problems. And so the call went out for military volunteers to serve in a civilian space agency. Characteristically, Glenn leaped at the opportunity.[5]

NASA officials established the criteria for selection in January 1959.[6] During that month, NASA reviewed 508 applications, narrowing the field to 110—5 marines, 47 navy fliers, and 58 air force pilots. Some rejected candidates were simply too tall (over five feet eleven) or too old (forty or older). Another requirement—a college degree—was waived in the cases of men like Glenn and Scott Carpenter, both of whom had undergone extensive military training after some time spent in college. (Carpenter was just one course short of graduation from the University of Colorado.) No army officers met the criteria because the army had no test pilots. That realization generated a futile mad dash to find acceptable army candidates.[7] NASA divided the remaining contenders into three groups and invited one cohort at a time to Washington for interviews and tests. Officials cut the pool to thirty-two through examinations as well as psychiatric, technical, and medical history interviews. NASA ordered those finalists to report to the Lovelace Clinic in Albuquerque for five days of rigorous testing.[8]

"They made every measurement you can possibly make on the human body, all the usual things you'd think about, plus all the other things that would occur in any natural physical exam and then things like, oh, cold water in the ear," Glenn recalled. Cold water was shot into the ear via syringe. The water and its temperature stirred up the natural fluids in the ear and caused the patient to experience dizziness and difficulty focusing his eyes. In analyzing a candidate's body density, doctors had them get in a water-filled tank, wipe off all bubbles on their bodies, and then submerge. Glenn wasn't sure how some of these exercises correlated to space travel.[9] It was easier to understand cardiologists' use of tilt tables to determine whether changes in orientation affected heart rate and blood pressure.

One of the men complained vehemently about his discomfort. Pete Conrad, who would join the astronaut corps in 1962, recalled vividly that after an enema, he was forced to walk down a hallway with a bag stuck in his bare rectum. As he hunched over, his lab gown hid nothing from curious passersby. He followed an orderly down one hallway, onto an elevator filled with people, and up another corridor before reaching a bathroom, where the clamped bag could be released and the toilet used. On the same day, as Tom Wolfe recounts in *The Right Stuff*, the doctors gave Conrad yet another request for an enema. Conrad marched into the office of the facility's director, General A. H. Schwichtenberg, dropped an enema bag on his desk, and said, "General Schwichtenberg, you're looking at a man who has given himself his last enema. If you want enemas from me, from now on, you can come get 'em yourself. You can take this bag and give it to a nurse and send her over—and let her do the honors. I've given myself my last enema. Either things shape up around here, or I ship out." Word of Conrad's protest spread quickly among the other candidates. The general was unapologetic, and Conrad stood firm.[10] (Conrad, as part of the second class of astronauts, later did multiple space walks and became the third man to walk on the moon. He commanded the first mission to the US space station *Skylab* and made a crucial space walk that saved the station.)

Ironically, NASA's approach to selecting astronauts was not terribly different from the one described in the first science fiction novel about a trip to the moon. "We do not admit sedentary, corpulent, or fastidious men into this retinue," Johannes Kepler wrote in *Somnium* (*The Dream*) in 1608, more than three centuries before the first lunar landing. "We choose rather those who spend their time persistently riding swift horses or who frequently sail to the Indies, accustomed to subsist on twice-baked bread, garlic, dried fish, and other unsavory dishes."[11] Long before spaceflight could be seriously contemplated, Kepler clearly expected spacemen to be daringly fast, like test pilots, and willing to endure discomfort for the privilege of leaving Earth behind, not unlike what astronauts and cosmonauts have endured since the first spaceflights.

Despite the depth and number of these tests, only one of the candidates—future Gemini and Apollo astronaut Jim Lovell—was disqualified in Albuquerque.[12] (Lovell was told that his bilirubin, a liver pigment, was high, but he passed his physical in 1962.)[13] The remaining thirty-one moved next to Wright-Patterson Air Force Base in Ohio. While the Lovelace exams had gauged overall health, tests at the Aeromedical Laboratory at Wright-Patterson measured the men's physical and psychological responses to stress caused by external forces such as heat, vibration, and acceleration. Decades later, Glenn remembered being placed in a heat chamber and the use of rectal thermometers. "I think it was 135 . . . and you had heat measurements being made on your body, including deep body temperatures," he said. "We stayed in there as long as you could until your pulse was too high."[14]

The thirty-six intelligence and psychological tests could be equally daunting. One asked the candidates to write twenty distinct answers to the question, "Who am I?" Glenn aced this exercise, writing, "I am a man, I am a Marine, I am a flyer, I am a husband, I am an officer, I am a father," and so on.[15] Another examination put each candidate in isolation for an undisclosed period of time while body sensors measured his responses. There was "no light, no sound, nothing," Glenn said. Candidates sat at a plain desk. Glenn ran his fingers over the desk and found what he was seeking—a drawer holding a tablet. He used a pencil from his pocket to scribble doggerel in the dark.[16] Another test asked would-be astronauts to answer the following questions as true or false:

1. I often worry about my health.
2. I am often unhappy.
3. Sometimes I feel like cursing.
4. Strangers keep trying to hurt me.

They also were given open-ended sentences to finish:

1. I am sorry that . . .
2. I can never . . .
3. I hope . . .
4. At times, . . .

And they received multiple-choice questions such as Light is to Dark as Pleasure is to (a) Picnic, (b) Day, (c) Pain, or (d) Night.[17] In another infamous test, candidates were handed a blank sheet of paper and asked what they saw. One saw "a blank sheet of paper." Another saw "a field of snow." Pete Conrad handed it back to his tester and reported, "It's upside-down."[18]

By March, NASA had cut the number of applicants to twenty-three. Slicing that number down to the anticipated six astronauts was a problem, so they chose seven—Scott Carpenter, Gordon Cooper, John Glenn, Gus Grissom, Alan Shepard, Deke Slayton, and Wally Schirra. They became the Mercury Seven, the men expected to make the first spaceflights. All were white males in their midthirties, self-identified Protestants, husbands, and fathers. (Few knew that Cooper and his wife, Trudy, were separated; Cooper realized that a reconciliation was a prerequisite for his career as an astronaut.)[19] Each man was endowed with notable intelligence and had been educated in public schools. All five college graduates had majored in engineering, as had Carpenter, who left college just short of graduation. In NASA's estimation, these seven displayed a clear understanding of the job's potential dangers without displaying signs of foolhardiness.[20]

NASA introduced the astronauts to the world at a press conference on April 9, 1959. They answered questions from a room full of excited journalists who treated them like a cross between the movie

At the news conference when the Mercury Seven astronauts were introduced to the public, all raise their hands in answer to a question about which ones were confident that they would return safely from outer space. From left to right, they are Deke Slayton, Alan Shepard, Wally Schirra, Gus Grissom, Glenn, Gordon Cooper, and Scott Carpenter. *National Aeronautics and Space Administration*

star John Wayne and the dashing aviator Charles Lindbergh. At first sight, reporters seemed to view the astronauts as heroic figures. "They're applauding us like we've already done something, like we were heroes or something," Slayton commented to Shepard,[21] and many Americans took their cues from the media. *Life* magazine, in an issue with a voluptuous Marilyn Monroe on the cover, described them as "smiling, clean-cut men."[22]

Glenn responded most naturally to reporters, telling them how much his loving family supported his decision to become an astronaut. "I don't think any of us could really go on with something like this if we didn't have pretty good backing at home," he said.[23] Slayton dared to disagree: "What I do is pretty much my business, profession-wise.

My wife goes along with it."[24] Author Matthew H. Hersch noted, "All but Glenn kept their answers formulaic."[25] When asked how they felt about their new roles, the seven pilots used the word "interesting" repeatedly, without expressing serious scientific curiosity. After a reporter asked them to describe the worst test they had undergone, Glenn stole the question, tailor-made for the jokester Schirra, and said, "That's a real tough one. It's rather difficult to pick one, because if you figure out how many openings there are on a human body and how far you can go in any one of them. . . . Now you answer which one would be the toughest for you."[26]

Faced with a question about religion, Glenn welcomed the journalists into his private life by telling them about teaching Sunday school. Most of his colleagues were reduced to mumbled answers. Carpenter described himself as "a faithful church-goer when it is possible." Grissom said, "I am a Protestant and belong to the Church of Christ. I am not real active in church." Slayton reported that he went to church "periodically." Shepard made a valiant attempt to return the conversation to space travel; nevertheless, under pressure, he revealed that he did not belong to a church but did attend a Christian Science church "regularly."[27]

Then the reporters' questions returned to astronauts' wives' opinions. Glenn quoted his wife as saying that he had "been out of this world for a long time," and he "might as well go on out."[28] His colleagues were flummoxed; these were military men, and all except one obviously felt uncomfortable discussing their feelings and beliefs. "We were surprised about John when we first saw John at the press conference, because we just weren't prepared for something like that," said Schirra. "And so John had the presence in front of a camera and in front of the press corps we didn't have."[29]

The astronauts quickly became human symbols of the US space program. During the week after the news conference, NASA headquarters received more than three hundred requests from journalists regarding access to the astronauts. To keep training on track and protect their families' privacy, NASA announced that the astronauts would not be available to the press anytime soon. The quiz show *I've*

Got a Secret put in a request for an astronaut, a wife, or a child to appear on the show. The astronauts would be unavailable on the date sought, NASA replied, while giving the astronauts permission to send wives or children if they wished.[30] (The parents of future moon walker Neil Armstrong appeared on the show in 1962 when Armstrong was chosen to join the second class of astronauts, known as the New Nine.)[31]

After naming the seven original astronauts, NASA issued a press release dividing astronaut training into six parts: education in the basic sciences, familiarization with the conditions of spaceflight, training in the operation of the Mercury space vehicle, participation in the vehicle development program, aviation flight training, and integration of the astronaut with ground support and launch crew operations. The vehicle development program assigned each astronaut to provide oversight for part of Project Mercury's planning.[32] Glenn was selected to guide decisions on the color-coded cockpit layout and worked closely with manufacturer McDonnell Aircraft Corporation to make the coding and placement of 165 cockpit gizmos—toggles, levers, lamps, dials, and meters—easy to use.[33] Slayton took responsibility for the Atlas rocket to be deployed in orbital flights; Cooper, for the Redstone rocket to be used in suborbital missions; Shepard, for flight tracking and space capsule recovery; Schirra, for life support in the pressure suit and in the capsule; Grissom, for automatic and manual control systems; and Carpenter, for communication and navigation.[34]

As the astronauts went to work, NASA could only hope that it had identified the best way to train men for a job that had never been done and would never be undertaken until these men did it. (A similar process was occurring in the Soviet Union, but Cold War secrecy made each nation begin with a blank sheet of paper and build a space program from there without having the benefit of consultation.)

In an interview on March 25, 1960, with three other astronauts—Carpenter, Grissom, and Slayton—Glenn again played the role of philosopher. Slayton called spaceflight a natural extension of a career as a pilot, while Glenn said, "Human beings I guess, have always wanted to explore. It's innate in all of us that we wish to explore as far as we can in our own way, whether it's Columbus coming to America or whatever type exploration it is, and we feel that this is probably the greatest exploration of all time." Carpenter tried to minimize the difference in responses by saying, "We all feel about the same about the whole project."[35] When asked about the risk, Glenn said, "We're working every day to make it as safe as we can, but we are all well aware that there are risks and will be right up to the time of firing. But if the stakes are high enough, you're willing to take those risks."[36]

During this interview, the astronauts talked about their united push for a window in the capsule—a battle they had won in time for it to be incorporated in the second manned spaceflight. However, Glenn was quick to say, "We're on the pinnacle of the publicity part of the program, but we certainly didn't design this thing ourselves; we don't want to give anybody that impression. This has been the result of much blood, sweat, and tears by many thousands of engineers all over the country to give us the capability of doing what we want to do. We certainly want to give them . . . some credit for this, too."[37]

In May 1959, George C. Guthrie, aeronautical research engineer, filed the first report on astronaut training to date. The seven astronauts' first lesson was on "NASA policies and organization." In their initial month, they had traveled to the McDonnell factory in Saint Louis, where the capsule was being built, as well as to Cape Canaveral, where it would be launched, and to Washington. They had listened to lectures and participated in computer simulations. The report notes that six astronauts watched an Atlas rocket liftoff at Cape Canaveral. Interestingly, it fails to mention that the rocket lit up, rose in the air, and then exploded startlingly above the heads of six would-be rocket riders.[38] After the explosion, Shepard said to Glenn, "Well, I'm glad they got that one out of the way. I sure hope they fix that."[39]

Over that summer, the astronauts faced a decision about how to handle the reporters, photographers, and cameramen who were surging into their lives at every opportunity. Glenn compared the attention to a tidal wave.[40] With the help of an agent who refused pay, they reached a deal with *Life* magazine, giving that periodical exclusive access to the astronauts' homes and their families. In return for this access, the astronauts would split $500,000, providing $24,000 a year for each man during the three years of Project Mercury. This was a windfall for men accustomed to earning $5,500 to $8,000 in annual base pay, plus about $2,000 in housing and subsistence allowances, and roughly $1,750 in extra flight pay. Articles would be written by *Life* staffers, although the magazine often indicated that an astronaut or an astronaut's wife authored a piece.[41] NASA approved the deal.

In training, the astronauts began experiencing what spaceflight might be like. Centrifuge exercises mimicked changes in the gravitational forces, or g-forces, that would be experienced in liftoff and reentry into Earth's atmosphere. After the seven began using the centrifuge and trying the rocket escape tower procedure in August 1959, Glenn, the eldest of the group, used the equipment more than any of his colleagues—twenty-nine times—and Shepard took part in exercises twenty-six times. In contrast, Schirra, who was tied up with laboratory work, had completed only twelve trials, and Slayton, who had a virus, participated just eleven times.[42] They used the centrifuge in what they called EIEO runs to simulate a splashdown. "Eyeballs in" was a normal liftoff or splashdown position with the astronaut on his back; "eyeballs out" simulated a splashdown with the cone facing down and the astronaut thrown against his straps.[43] They also experienced brief spurts of weightlessness, floating through the air on large planes flying parabolic courses.

During the six months between July 1 and December 31, 1959, a typical astronaut traveled one out of three days. Some trips were individual, allowing each astronaut to explore his specific area of responsibility for Project Mercury. They also trekked for "fittings"—for pressure suits or for spaceflight couches that would be made to fit each man's body exactly. They visited rocket designer Wernher von Braun's facilities in Huntsville, Alabama; they learned about star patterns at the University of North Carolina planetarium in Chapel Hill; and they toured the factory where the Atlas rocket was being produced in San Diego. During the summer of 1959, the astronauts got their first chance to try out something close to Mercury controls at Langley Research Center in Virginia, where NASA management was based at that time. An orbital simulator allowed them to practice working with attitude controls and retro-rocket fire. They also learned about astrophysics, meteorology, geophysics, trajectories, and much more. To anticipate the possibility of a failure in environmental systems, they underwent three hours in a sealed chamber, and none responded badly to the lack of fresh air. Most of what they did on Earth was keyed to their space missions. Earthbound medical readings would serve as baselines for comparison during flight, science lessons would prepare them to be good observers, and training in the use of scientific equipment would sharpen their findings.[44]

The astronauts worked together, and yet they were aggressively competitive. Each wanted to be the first man in space, a prize for which they could not compete openly. Consequently, they often vented their energy in other ways—illegal car races, waterskiing, climbing, and just about any other daring activity in which one person could show superiority over another. These contests were taken so seriously that Cooper waited more than thirty years to admit that he had secretly altered his Corvette to beat Shepard in one race. His confession came after Shepard's death in 1998.[45]

Kidding each other became part of their daily routine too. Through the course of training, Carpenter and Glenn became friends, and decades later, Carpenter remembered the two astronauts were heading to the airport in a convertible for a joint trip—and they were late. Carpenter had their airplane tickets in his hand, and as a joke, he released the envelopes and told Glenn that the tickets had flown out of his hand. Glenn "laughed about it and [said] we'd take another plane." Carpenter noticed that after he had revealed the joke, "his laugh had a different note."[46]

Over time, most of the other astronauts became irritated by Glenn's perfectly choreographed public appearances, his by-the-book behavior, and what they viewed as a "holier than thou" attitude. He did not share his colleagues' love of shiny new sports cars, and they didn't appreciate what they considered to be his overly passionate public embrace of God and country. During astronaut training, Glenn decided to spend Monday through Friday at Langley and to visit his family only on the weekends. He established a daily habit of running two miles before breakfast, reducing his weight from 195 pounds to 168.[47] The other astronauts knew what Tom Wolfe later put into words: "Among the seven instant heroes, John Glenn's light shone brightest."[48] He was too perfect for his fellow daredevils. They called him "the Boy Scout."

Things came to a head when the astronauts were staying at the Kona Kai Hotel on Shelter Island in California. They were there to meet with Atlas engineers. One night, Glenn received a middle-of-the-night call from John "Shorty" Powers, the NASA press officer. Powers told him that a newspaper had called to say that it had photographs of an astronaut with a woman other than his wife and planned to publish them. Glenn called the newspaper's publisher and managed to convince him that releasing the photos would provide a victory for "godless communists."[49] He wrote in his memoir, "I pulled out all the stops."[50]

The following morning, Glenn called for a "séance"—a meeting of the astronauts only, usually reserved for discussions of engineering decisions they wanted to challenge. Glenn told the others that any astronaut who had difficulty keeping his "pants zipped" was a threat to the space program. Shepard angrily countered Glenn's argument, saying that their personal lives were not the press's business or Glenn's. Like it or not, Glenn replied, the astronauts' behavior was being scrutinized. Most of the astronauts sided with Shepard then and later reluctantly conceded Glenn's point.[51] At this point, Glenn, who never identified the womanizing astronaut, didn't care what his colleagues thought of him and made no effort to avoid alienating them.

In July 1960 the astronauts underwent desert survival training near Reno. Given the possibility that a capsule might land in a North African or Australian desert, this exercise was intended to teach astronauts what they needed to know about survival in an arid environment and to evaluate the adequacy of Mercury survival kits.[52] In classroom training, the astronauts had been taught the best survival techniques, how to prepare a shelter, and how to recognize the symptoms of dehydration. Then they were dispersed to remote areas so that each man was on his own for seventy-two hours. Every astronaut began with six pints of drinking water—the amount that would be available in a capsule. On subsequent days, water was replenished. While Slayton later reported following the instructions to the letter and finding them quite effective,[53] Glenn, in his continuing quest for new experiences, uncharacteristically decided to ignore the directions he had been given. He wanted to undergo dehydration, so he did not consume water in small amounts as recommended. Instead, he ate and drank nothing in the first twenty-four hours and found himself significantly fatigued. He then began consuming water "in as large quantities as I desired." He felt much better after the first consumption of water and considered himself healed after a nine-hour period when he consumed fifteen pints of water. (The man assigned to check on each astronaut gave Glenn an extra supply.)[54] As his organs responded to the influx of water, he felt his body temperature and respiration drop. He was surprised by the dramatic effects of dehydration and said he would be able to identify them easily in the future.[55]

Shortly before Christmas 1960, Bob Gilruth, who headed the manned space program, called the astronauts together and asked each man to take a "peer vote" on whom he would like to see get the first spaceflight if he did not. Glenn was flabbergasted that the astronauts' opinions were a factor. He had worked to earn the respect of his superiors and had made no effort to win a popularity contest among his colleagues.[56] On January 19, 1961, the day before JFK's presidential inauguration, Gilruth told the astronauts who would get the first spaceflights: Shepard was to take the first mission, with Grissom flying second, and Glenn acting as backup for both.

Shepard could not hide his jubilation. "There I am looking at six faces looking at me and feeling, of course, totally elated that I had won the competition."[57] Glenn was first among the other astronauts to step forward and shake Shepard's hand; nonetheless, he subsequently did his best to convince NASA that both the peer review and the selection of Shepard had been mistakes. Officials replied that Shepard had performed better on the procedures trainer than Glenn, and they eventually told Glenn bluntly to stop complaining and accept the verdict. Shepard's biographer Neal Thompson has speculated that Glenn's image consciousness led "some NASA officials to consider Shepard the better choice for the first flight."[58]

NASA decided not to announce Shepard's selection, but the agency did reveal that Shepard, Grissom, and Glenn were the three top contenders. Those three appeared on *Life's* cover in February, and it was hard to miss the fact that naming them "the first three" unavoidably cast Carpenter, Slayton, Schirra, and Cooper as "the other four," a demotion that did not thrill them.[59] Schirra said those not among the top three "were devastated."[60] At the same time, neither Grissom nor Glenn appreciated being forced to act as if he still had a chance of making the first flight when he did not.[61]

As Shepard prepared for his flight and the other astronauts struggled to accept their new places in the lineup, some experts continued to argue that men did not have to climb aboard rockets at all; machines could do the work, they contended. Conversely, a report prepared for Vice President Lyndon Johnson in 1961 proclaimed, "Sending only instruments to probe the mysteries of space will not provide man with the ultimate knowledge that he must have to assess the values that space exploration may bring in man's drive to break out into the universe."[62] Moreover, the report argued, "The world effect of landing men on other bodies in space and returning them will count for much more than instruments however ingenious."[63] Like archaeologists in reverse, astronauts had to go out into space and explore other heavenly bodies to glimpse the future, and that vision had the potential to change the world.[64]

On April 3, Edward R. Murrow, director of the US Information Agency, wrote a memorandum to the White House proposing a possible US reaction to a Soviet failure that cost the life of a cosmonaut—a recommendation that was not needed.[65] Shepard lost the opportunity to be the first man in space on April 12, when Yuri Gagarin of the Soviet Union grasped that honor—and Gagarin successfully orbited Earth once rather than making a fifteen-minute up-and-down suborbital flight like the one planned for Shepard.

During training for Shepard's flight, he and Glenn minimized past disputes and worked together closely, spending long hours in simulators.[66] The two shared frustration when rain canceled Shepard's scheduled May 2 liftoff. NASA had planned to reveal the first astronaut immediately before liftoff; nevertheless, after the rainout, the agency announced that Shepard would take the first flight. Two days later, Shepard and Glenn took a jog together in a beach area near the launchpad and did not dwell on the forthcoming day's scheduled liftoff. That night, all seven astronauts shared a roast beef dinner. Shepard, who loved teasing Glenn by calling him "my backup," toasted him and said, "John's been most kind."[67]

On May 5, Shepard arose shortly after midnight to prepare for the flight of *Freedom 7*.[68] He had intended to speak to reporters before riding the elevator seventy feet up to his capsule atop a Redstone rocket; however, when the moment came, he found he was too overwhelmed with emotion to follow through on that plan. Glenn, as his backup, was with him when he entered the capsule. He had left two jokes for Shepard—a sign that said No HANDBALL PLAYING IN THIS AREA and a centerfold from a girlie magazine. When Shepard saw Glenn's smiling face, he thought, "I'll be damned. He's becoming a damned prankster." Glenn then buried that thought by immediately reaching forward and removing both items so that they would not be caught in photos of the capsule's interior. As technicians prepared to close the hatch, Glenn stuck his hand out and shook Shepard's. "Happy landings, commander," he said.[69]

During the countdown, there was a problem with the inverter, a box that changed DC electrical current into AC electrical current in the Redstone, so NASA decided to pull the gantry back and replace the inverter. Shepard asked Cooper, the voice communicator in the Cape Canaveral block house, "Gordo, would you check and see if I can get out and relieve myself quickly?" After a few minutes, Cooper responded in an imitation of von Braun's German accent, saying, "The astronaut shall stay in the nose cone." He was advised to unplug the wiring in his suit and relieve himself within the suit. Then he had to lie there wet and wait for liftoff. Shepard reported that the suit was dry by the time he launched.[70] After fifteen minutes and twenty seconds in flight, Shepard splashed down in the Atlantic Ocean. He had reached an altitude of 116.5 statute miles before the forces of gravity drew him back to Earth.

Three days later, all seven astronauts went to the White House, where Shepard received the NASA Distinguished Service Medal for what President Kennedy called "an overwhelming success." While reading the commendation, Kennedy dropped the medal and quickly retrieved it before anyone else could. He gave it to Shepard, calling it a "decoration which has gone from the ground up."[71] Afterward, the astronauts went to Capitol Hill, and there was a parade, with Shepard riding alongside Vice President Lyndon B. Johnson in the lead car. Shepard was unhappy about the way his fellow astronauts were treated. "I hated their having to ride along in my wake, staring at the back of my head," he commented. "I was shocked at one point to look behind, and there were Deke Slayton and John Glenn trying to hitch a ride in a news media pool car." The journalists did not recognize the astronauts and physically pushed them away until someone from NASA identified them.[72] Shepard's time atop the astronaut heap would be short lived, but no one foresaw that anyone could surpass the popularity of the first American in space.

In Washington, the Kennedy administration was doing its part to accelerate the space race. The president and his team were struggling to recover from international embarrassment. Gagarin's flight had shattered hopes of sending the first man into space, and the administration had made a disastrous mistake in supporting an unsuccessful exile-led invasion of Cuba at the Bay of Pigs on April 17.

A surprisingly prescient Bureau of the Budget report issued on the day of Shepard's flight established goals for the space program: to make scientific discoveries, to gain a military advantage, to improve weather forecasting, to encourage partnerships with other nations, to bolster the development of new technologies, to put men in space, and to expand and improve communications capabilities for civilians as well as the military. Much of this work would be achieved via unmanned satellites.[73] Three days later, JFK received a report from Secretary of Defense Robert McNamara and NASA director James E. Webb declaring that the United States was ahead of the Soviet Union in using the space program to gather military and scientific information as well as in identifying the commercial/civilian uses of space vehicles. However, the report conceded that the Soviet Union had bested the United States in accomplishing "space spectaculars which bestow great esteem."[74]

In a May 25, 1961, address to a joint session of Congress, Kennedy raised the stakes: "I believe that this nation should commit itself to achieving the goal, before this decade is out, of landing a man on the moon and returning him safely to the earth. No single space project in this period will be more impressive to mankind, or more important for the long-range exploration of space; and none will be so difficult or expensive to accomplish."[75]

Project Mercury occurred at the time that many have called "the coldest days of the Cold War." More than once, the United States and the Soviet Union appeared to be on the brink of nuclear war, most notably during the October 1962 Cuban Missile Crisis. Behind closed doors, cautious leadership on both sides avoided stumbling into war; in public, fierce rhetoric remained the order of the day. Cold War stress and competition added momentum to the space race—and reduced quibbling on Capitol Hill about the expense of the program. Beating the Soviet Union was the bottom line for many members of Congress, and they were willing to spend extravagantly to reach that goal. Kennedy had difficulty passing much of his domestic legislative program, but when spending related to the ongoing rivalry with the Soviet Union, he could count on congressional support. The rush-to-the-moon venture was a potentially winnable contest, and the cost of the project would not hamper its success. The Soviet Union had succeeded in scoring a number of space firsts, but sending men to the moon and returning them safely to Earth would require a great deal of sustained spending to develop the equipment and the human proficiency necessary to make it happen. Kennedy had made his challenge, and he would continue pushing the United States to shoot for the moon.

In July, Gus Grissom followed Shepard into space. While the parameters of his suborbital mission were similar to Shepard's, there were differences. Grissom's capsule, *Liberty Bell 7*, had a different design that incorporated a window, which had been sought by the astronauts, and a new hatch. After a successful suborbital flight and splashdown, the hatch blew open. *Liberty Bell 7* began to fill with water. Grissom evacuated and almost drowned in choppy water before divers could reach him. The capsule sank to the ocean floor—a big loss for NASA.[76]

Shortly after Grissom's flight, Gherman Titov of the Soviet Union orbited Earth seventeen times, making US advances seem small. An August 1961 report from the Space Science Board of the National Academy of Sciences backed the moon goal, saying, "The primary scientific aims of this program are immense: a better understanding of the origins of the solar system and the universe, the investigation of the existence of life on other planets and, potentially, an understanding of the origin of life itself."[77]

Both Titov's ambitious flight and the president's determination to reach the moon by the end of the decade affected NASA's thinking about Project Mercury. The next anticipated flight was another suborbital mission by Glenn. "John might have been unhappy [with being the third astronaut], but he was still resourceful," Schirra later wrote. "He was soon saying that we really didn't need a third suborbital flight."[78] Walt Williams, NASA deputy associate administrator at this time, said, "None of us really thought a third sub-orbital flight was necessary. Any time and energy spent on them didn't get used on the Atlas, where it was needed most." Williams feared that Glenn would be upset about cancellation of his suborbital shot and the delay of his first spaceflight until "it turned out he had his sights set on the first Atlas orbital mission."[79]

On October 4, 1961, Gilruth told the Mercury astronauts that Glenn would make the first orbital flight, with Carpenter acting as his backup. A successful November 29 flight by a chimp named Enos cleared the way for the Atlas rocket to carry a man into space. Weeks of postponement lay ahead; yet Glenn was headed into space on a flight that would captivate the nation and the world.

President Kennedy often emphasized the openness of US spaceflights as opposed to secretive Soviet flights made public only after the cosmonauts' safe return to Earth. Kennedy realized this was a gamble. If a space disaster had happened, Americans might see it with their own eyes, as they did more than twenty years later when the space shuttle *Challenger* suffered a catastrophic accident after liftoff. More basically, however, Kennedy may not have fully anticipated the public relations advantage of US flights that attracted worldwide attention as they transpired in plain view. John Glenn would be the first to take humans from around the globe on a voyage into space.

5

AROUND THE WORLD
IN 89 MINUTES

John Glenn awoke at 1:30 AM on February 20, 1962, ready to explore the mysteries of orbital spaceflight—and there were many unknowns to confront. Could a human swallow food in a weightless environment? Would his inner ear be affected, making him lose his equilibrium? Would his eyes change shape? His vision blur? Would weightlessness become a rapturous addiction that he did not want to leave behind? All these questions and more waited for Glenn to cast them aside and replace them with facts.

Soviet engineers and scientists knew the answers to most of NASA's questions. However, the Soviet space program was shrouded in secrecy, and few facts slipped beneath the Iron Curtain that separated the United States from its Cold War adversary. The Soviet Union's most recent cosmonaut, Gherman Titov, reportedly suffered unrelenting motion sickness throughout his flight. Neither Glenn nor

NASA could be sure whether that was an inevitable part of orbital space travel or a problem specific to Titov.

The United States was playing catch-up. The Soviet Union had considerably more knowledge about orbital flight even before Yuri Gagarin's first manned flight. At that point, Vostok vehicles, including many carrying dogs, had totaled about one hundred orbits. Before Glenn went up, the United States had finished only three orbits of Mercury spacecraft—one with an empty capsule and two carrying the chimp, Enos.[1] This lack of experience compounded the ignorance of American scientists.

The time had arrived to begin finding answers. On the morning of his flight, Glenn awoke in the top bunk of crew quarters and saw the face of physician Bill Douglas, who was leaning on the edge of his bunk. The eager astronaut arose, showered, shaved, ate steak and scrambled eggs for breakfast, underwent a medical examination, and slithered into his space suit. Meanwhile, Scott Carpenter, Glenn's backup, checked the capsule to guarantee everything was working properly.

As he moved from one preflight task to the next, Glenn would have been hard pressed to avoid a sense of déjà vu. He had done this before—more than once—and had waited for liftoffs that never happened. Feeling playful on this day full of promise, Glenn played a trick on his doctor. To confirm that the air in the suit was pristine, Douglas always ran a hose from the suit into a fishbowl filled with tropical fish. As he performed what had become a routine procedure, Glenn asked Douglas whether he had noticed that two of the fish were floating belly up. Douglas raced to the fish bowl, only to realize that he had fallen prey to Glenn's clean and sober sense of humor.[2]

Since Glenn had first readied himself for flight in late 1961, ten scheduled launch dates had come and gone, mostly as a result of weather problems at Cape Canaveral or in the capsule's splashdown recovery area. He spent most of January at Cape Canaveral, while his family remained at home in Arlington, Virginia. While living in crew quarters, his only regular outings were attendance at church on Sundays and trips to Bob's Barber Shop in Cocoa Beach every few days so that the tight fit of his custom-made helmet would be just right without a single errant hair on the astronaut's balding head. His church, his job, and his family were the pillars of Glenn's life, and he found little time for extracurricular activities—even amid what seemed like an unending series of delays. Throughout the wait, Glenn stayed focused on the launch of the MA-6 mission in his capsule, *Friendship 7*, which had been named by his children. Much to the other astronauts' amusement, he had requisitioned an artist to hand paint the name on his ship.[3]

The first potential launch dates occurred in December 1961. After various issues made the mission impossible, administrators looked at the calendar and decided that delaying the launch until after the holidays would be kindest to NASA employees and others working on the project. As the weeks passed, impatience gained a stranglehold on journalists and space fans who hoped to see the liftoff from the sands of Cocoa Beach. Curious and restless reporters asked psychiatrists whether Glenn could stand the strain of so many postponements, but the astronaut kept smiling. The delays gave him more time to perfect his mission, he told journalists. Indeed, Glenn logged almost sixty hours of training on the procedures trainer between December 13 and February 17, roughly twice the amount required by NASA directives. He also participated in a two-day recovery exercise in December, and during the month before the flight, Glenn underwent two "pad rehearsals" and one simulation of his entire mission.[4]

As he said, he was prepared, but his apparently unending patience was an act. He had sat on top of that rocket on three occasions without launching, and he admitted decades afterward that each delay triggered a huge letdown. "When you go through all that and get ready and get yourself psyched up to go and all ready, then they cancel the flight and you have to get unstrapped and get out and come down and start all over again—a big disappointment," he said. "So I hated to see a flight canceled."[5]

While he was eager to lift off, Glenn was not worried about the Soviet Union charging further ahead in the competition. He spurned the idea of the space race, arguing, "We are not in a drag race with the Russians. . . . We won't go until ready whether the Russians go tomorrow, the next day, or a day ahead of us or a day after us. We will go when our project is good and ready to go." This belief, solidly held since day one in Project Mercury, went a long way to explain his confidence in NASA and his willingness to wait as long as there were good reasons to do so.[6] He knew there were real dangers even on days that seemed perfect for liftoff. As NASA administrator James Webb told a radio interviewer during the delay-filled January 1962, "There are always items of physical danger in these highly experimental flights—when they're in the air or out in space beyond the air. The amount of danger, we believe, is equivalent to the test pilot flying the first flight of an experimental airplane."[7] Doing everything possible to ensure the astronaut's safety was the main reason for postponements.

Glenn himself came close to adding to the delays or costing himself the flight. In January, which was very late in the game, he allied himself with Project Mercury's training officer Robert Voas in trying to change the flight plan. They appealed to flight director Chris Kraft, saying they worried about the prospect that Glenn might have to handle retro-rocket firing on manual and might have a problem. Kraft was puzzled; retrofire was not manual unless there was a problem with the automatic system, and Glenn had tested the procedure on a simulator and was "pretty good at it, too." It was impossible to do more advanced training unless the ship was in orbit. Voas explained that they were concerned the capsule might start spinning and that Glenn would have trouble setting the retrofire under those conditions. Consequently, he and Glenn endorsed the idea of intentionally creating a spin on all three axes during the first orbit so that he could experience having to bring the capsule back under control. Kraft was outraged. "So you want to create an actual emergency situation in space just so he can practice recovering from it, in case the same emergency happens by accident?" Kraft flatly rejected the suggestion, saying Glenn had enough to do without adding a "dumb idea."[8] No changes would be made to his mission.

Still, no one could say for sure when Glenn's flight would launch. In another threat to his quest, he was exposed to children with the mumps, which could have awarded his flight to Scott Carpenter. Fortunately, he did not contract the childhood malady.[9]

On January 27, Glenn climbed into the capsule, only to lie on his back for about five hours before NASA decided bad weather made a liftoff impossible. NASA's analysis of the astronaut's fatigue had been a factor in the decision to give up after so many hours of waiting.[10] At the Glenns' home, Annie had received a supportive phone call early in the morning from Vice President Lyndon Johnson, along with his wife, Lady Bird.[11] Now Johnson wanted to visit her at home and show his sympathy for her stressful day.

Because of her stuttering problem, Annie Glenn was not eager to invite Johnson and the reporters following him into her home.[12] Much to Johnson's surprise, she refused to see him. Loudon Wainwright of *Life* was at her home, and she felt comfortable with him. Rather than reveal her true reason for rejecting Johnson, the Glenns later said that if Johnson came in, Wainwright had to leave so that other reporters could enter with the vice president. This would have violated the astronauts' deal allowing only *Life* staffers into their homes. Annie felt comfortable with the *Life* reporters who had become regular visitors to her home. However, when she had to talk to a hoard of other reporters, she limited her speech to short sentences to avoid putting a spotlight on her stuttering. Now she confronted an uncomfortable fact: Lyndon Johnson did not accept defeat easily.

Therefore, NASA administrators summoned Glenn to a conference room minutes after he emerged from the capsule. The tired and discouraged astronaut arrived in a bathrobe, having just stepped out of his space suit. His supervisors urged him to change Annie's mind. If he did not act, they threatened to replace him on the flight, Glenn later said. Unfortunately for them, he was feeling no more flexible than Johnson. He remembered saying, "You call your press conference to announce that, and I'll call my press conference in rebuttal, and we'll see who comes out best." More than forty years later, as he looked back on it, he was confident that NASA never intended to take his flight away from him.[13]

In the days following the latest aborted liftoff, a routine check revealed fuel between bulkheads of the enormous Atlas rocket. Fuel had leaked into a small cavity between a layer of insulation and a bulkhead. Correcting this problem, the launch vehicle team projected, would require at least ten workdays. Such a long postponement would affect the recovery operation, involving sixty aircraft, twenty-four ships, and eighteen thousand military personnel. The navy faced decisions about whether ships, planes, and their crews could stay in recovery zone positions or whether replacements would have to be deployed. Trying to work quickly, NASA technicians removed a bulkhead to eliminate the space that made room for a leak.[14] The soonest possible launch date became February 13.

Issues interceded, and the thirteenth day of February was not Glenn's launch date. Nevertheless, NASA officials called a press conference on that day to talk about the flight, the problems preventing a liftoff, and the general state of Lieutenant Colonel John Glenn. Out of this conference came one of the clearest analyses of the flight's importance. Air force major Charles Gandy said, "On this flight, we must prove that the machine works and that man can work with it."[15] Glenn was ready for the challenge.

On February 20, he was optimistic: this might be the day. The weather forecast for the recovery zone was good, and meteorologists believed cloud cover at Cape Canaveral might break at midmorning. After suiting up, he stepped aboard a transfer van that took him to the launchpad. When he arrived, he stayed in the van through two holds in the countdown—one for weather and one related to the Atlas guidance system. Around 6:00 AM, he emerged, smiling and waving at the cameras. And he greeted the people laboring around the launchpad, workers who shared his mental fatigue surrounding the many delays. He gazed up at the Atlas rocket, noting that the bright lights illuminating it had created an eerie bluish glow.

More holds followed. A microphone on his helmet broke and had to be replaced. "I guess we have probably pushed those mikes up and down many thousand times with no trouble, and it had to break right in the middle of the count," Glenn remarked later.[16] And when the crew closed the hatch, a bolt came apart. Just replacing one bolt devoured another forty minutes. Gus Grissom had flown with a

Glenn heads to the Atlas rocket that would carry him into space.
National Aeronautics and Space Administration

broken bolt on the hatch. Notwithstanding that precedent, this bolt had to be replaced.[17] It was Grissom's flight that suffered some kind of post-splashdown problem involving the hatch, which may have elevated anxiety over this issue.

Glenn knew vital minutes were slipping away. If mission control determined there was not enough time for Glenn to complete his flight and provide adequate daylight hours for recovery in the Atlantic, his voyage would be shortened or postponed yet again.

Glenn became edgy as flight controllers asked him repeatedly to check switches and set fuses. After the flight, he complained that it seemed like make-work. He described it as "a sort of WPA program designed to keep the astronaut busy." He found it unnecessary. "Let the fellow in the capsule have as much spare time as you can," he later advised NASA.[18]

Nonetheless, the space agency was relentless about readiness. NASA had prepared for both unexpected and undesirable outcomes. NASA had written a script for an astronaut to deliver in case he landed in another country. The message, written out in five languages—English, French, Spanish, Hausa, and Swahili—said:

I am an American. I have returned to the earth in this vehicle after a test flight in space for peaceful purposes. I ask your help. Please tell your officials that I am here. Please inform the nearest American Government officials that I am here. I will stay with this vehicle until arrangements can be made to return me and the vehicle to my country. Thank you.[19]

The text offered a slight twist on the old science-fiction movie line: "Take me to your leader." NASA also had prepared statements for the president, the vice president, the NASA administrator, and Alan Shepard to read if Glenn died during the mission. Here's the script written for President Kennedy:

To Mrs. Glenn and members of the Glenn family go my deepest sympathy. It was my pleasure to have known John Glenn. This nation and the entire world share his loss with the Glenn family. Space scientists will revere his pioneering spirit.

Johnson's speech proposed a permanent scholarship fund in Glenn's

name, and the flight director's statement acknowledged how dangerous the "flight development field is."[20]

On this day, as the clouds scattered, giving way to clear skies,
Glenn knew a liftoff was likely—after a hold already underway to
replace a frozen fuel valve. He asked Carpenter to put in a call to his
Virginia home, and he repeated the message to Annie that he had
twice delivered on his way to war: "Don't be scared. Remember, I'm
just going down to the corner store to get a pack of gum."[21]

Despite this lighthearted repartee with his wife, he later admitted
to feeling odd sensations during the hours spent sitting on a rocket.
When he moved, slightly repositioning himself on the couch, he could
feel the rocket sway as liquid oxygen fuel responded to his motion.[22]
The count picked up again at 9:25 AM, only to be derailed one last time
when a power failure in Bermuda threatened to make a key tracking
station inoperable. That issue was resolved within a few minutes, and
the countdown commenced, moving steadily toward zero.

In the final seconds, Scott Carpenter famously said, "Godspeed,
John Glenn."[23] Millions of earthbound humans later heard it, but in
that moment, John Glenn did not, because Carpenter was not on the
correct frequency.

Through the entire wait, Glenn's pulse had remained 70–80. In *The
Right Stuff*, Tom Wolfe compared that to the pulse rate of "any normal
healthy bored man having breakfast in the kitchen." Even when the
countdown reached its end, his pulse rate rose to just 110.[24] (The target
heartbeat for a forty-year-old is 90–153.)

At 9:47 AM, three hours and forty-five minutes after Glenn had
entered the capsule, the world around him suddenly vibrated violently
as he sat atop a growling giant that had just come to life. He felt the
rocket's burst of energy before he heard its roar. Beneath him was a force
as powerful as the thrust of sixteen hundred steam locomotives. Calmly,
Glenn's voice reported, "The clock is operating. We are underway."[25]

The Atlas had been problematic in early tests; its thin, flexible skin
could not stand up to the period of maximum pressure at twenty-five
thousand to twenty-six thousand feet in altitude in what is called the
Max Q area, the point at which the rocket encounters the greatest

stress as it passes through the atmosphere. With time, engineers found a solution: a heavy stainless steel "belly band" to help the booster harden its shell and pass through the danger zone without difficulty. Problem solved, but Glenn could not be expected to forget the Atlas rocket explosion he had watched in 1959. And now he was ascending on an Atlas with a belly band. Destination: outer space.

The growing g-forces crushed him against his seat. After forty-eight seconds, he reached the rattling pressures of Max Q, and by one minute, twelve seconds, the ride became smoother. Less than a minute later, the booster engine shut down, and soon afterward the astronauts' launch escape tower rode its own rocket into oblivion. The sustainer engine and two smaller Vernier engines pushed the capsule ever higher. As Glenn again felt g-forces mounting on his body, the decision to attempt orbit was made. "Cape is go and I am go," he said. Soon, Shepard, the capsule communicator at Cape Canaveral, notified him that the sustainer engine would shut down within seconds.[26] The Atlas had been "turning a corner in the sky," in Glenn's words.[27]

Just five minutes and one second after liftoff, powered flight ended normally, with the sustainer engines expiring.[28] Next, Glenn felt a jolt and heard a loud noise as posigrade rockets fired, breaking the connection between the capsule and the Atlas. Seconds later, Glenn reported, "Zero-g and I feel fine."[29] He was gliding into the proper orbital position.

NASA announced the news: "The *Friendship 7* spacecraft is in orbit with John Glenn at the controls." His speed was 17,530 miles per hour—more than three times as fast as Alan Shepard had traveled. The apogee, or high point of his orbit, was 160 miles, and the perigee, or low point, was 100 miles.[30] What NASA did not announce was that the posigrade rockets had fired 2.5 seconds late. That caused an error that had to be corrected by the capsule's attitude control system, which determined its orientation. As a result, the process of positioning the capsule correctly into orbit unexpectedly used 5.4 pounds of the craft's sixty-pound fuel supply.[31]

As Glenn looked out the window, somewhat dirtied when the tower was jettisoned, he described the view as "tremendous" and "a

The Atlas rocket lifts off with Glenn in the *Friendship 7* capsule on February 20, 1962. *National Aeronautics and Space Administration*

beautiful sight."[32] His 1.5-ton craft became the sixty-ninth American-made Earth satellite just four years after the first successful launch of a US satellite.[33] Almost immediately, Shepard at Cape Canaveral delivered good news: "You have a go, at least seven orbits."[34] Those words confirmed for Glenn that he was safely in space and would remain there for the duration of his planned three-orbit mission, but if needed the capsule would return to Earth on its own after circling the planet seven times. Still, the desire was to limit the flight to three orbits so Glenn would splash down in daylight, enabling navy ships in the recovery zone to more easily find him.

Each orbit would take eighty-eight minutes and twenty-nine seconds.[35] Almost before Glenn knew it, he was moving out of Cape Canaveral's communication range and on his way to his next connection, in Bermuda. Glenn's back was pointed in the direction he was moving, so Earth beneath him revealed itself after he passed over it. While he was in contact with the Canary Islands, he did a manual check of the ship's attitude controls. He was happy with their performance. Manual controls would be used only if the automatic systems failed.

One of his many assignments was to track the now-useless booster in its own orbit for as long as possible. He lost sight of it over Africa.[36] Tracking the booster provided evidence that would be useful in planning rendezvous missions that were an essential part of preparing for a voyage to the moon. Those missions would occur in the mid-1960s during Project Gemini, the second phase of the space program that would prepare the way to take Americans to the moon. Another *Friendship 7* experiment did not work at all. NASA had crafted a patch for Glenn to wear over one eye as a way of adapting to the day-to-night light changes; regrettably, the patch became covered with dust and would not stay in place. Consequently, Glenn stopped trying to use it.[37]

The astronauts had been encouraged to talk to their communications contacts about what they saw as well as what they were doing. It was a good tactic for maintaining a historic record, and it helped pilots to avoid accidentally tumbling into saying something inappropriate for the world to hear,[38] so when he talked to the capcom in Kano, Nigeria, Glenn said he could see what he thought were fires with long smoke trails right on the edge of the desert. The capcom replied, "A lot of this part of Africa is covered with dust."[39] The *Nigerian Morning Post* reported on its front page that Glenn had flown above Kano.[40]

Weightlessness fascinated and delighted Glenn. Even when he made swift moves in a conscious attempt to cause dizziness or queasiness, his body had no reaction.[41] No bad effects struck the same man who had been incapacitated by motion sickness on the ship taking him to his first World War II assignment in the Pacific. Doctors had worried that weightlessness might have an odd effect on the fluid in his inner ear, causing vertigo, so NASA had provided a syringe of Tigan, a motion sickness medication; however, Glenn didn't need it. He felt good.[42] Ophthalmologists had feared weightlessness's effects on the shape of the eye and his vision, so he had to check his eyesight every twenty minutes on a miniature eye chart.[43]

By 10:28 AM, forty-one minutes after liftoff, he was on the night-side of the planet, in darkness over the Indian Ocean.[44] He told the Indian Ocean capcom about the beauty of the sunset—his first of four on this day. As he neared Muchea, Australia, Glenn heard the familiar voice of fellow astronaut Gordon Cooper and greeted him by asking, "How are you doing, Gordo?" He and Cooper talked about his weather and star observations, and Glenn told Cooper, "That was the shortest day I've ever run into."[45] Then Glenn saw "a big pattern of lights" on the ground. Cooper explained that residents of Perth had left their lights on for Glenn in one sign of the flight's worldwide impact.[46] Glenn called the lights of Perth one of the greatest thrills of the trip. He learned that the city had turned on "every streetlight and neon sign they could get their hands on. Many Perthians rigged up homemade reflectors, and others spread out bedsheets to throw up more light."[47] A week afterward, the Oklahoma Gas and Electric Company published a thank-you advertisement honoring "our good friends of Perth." The ad was published in the *West Australian*.[48]

As he flew past Australia, sunrise occurred, and the new day delivered a surprise: he saw what looked like thousands of fireflies all around his ship. At first he believed that his craft had tumbled and that he was seeing stars. Then he realized these tiny visitors represented a new phenomenon. "I would say most of the particles were similar to first magnitude [very bright] stars," Glenn later wrote. "They were floating in space at approximately my speed."[49] As he rounded Earth three times, he saw "fireflies" at each sunrise. On the third orbit, he turned to face them and later commented, "Facing the direction of travel is, by the way, the best way to travel."[50] Glenn was mesmerized by these tiny fellow voyagers. He spoke so much about them that they became known as the Glenn Effect. He was convinced they were not a by-product of his spacecraft, and after the flight, he wrote an article for *Science* magazine that prominently showcased these mysterious strangers.[51] He took some kidding about the "fireflies" from NASA colleagues.

A few months down the road, when Scott Carpenter orbited Earth three times himself, he saw the Glenn Effect as well. "I have a delightful report for one John Glenn. I do see fireflies," he announced.[52] Carpenter realized that movement inside the capsule generated more of these itinerants, so he banged the capsule's hatch and watched them multiply. He believed the "fireflies" were crystallized water vapor that materialized when an astronaut's body moisture was released.

NASA scientist John O'Keefe later made a comprehensive study of the "Glenn Effect" and concluded that the miniature space travelers undoubtedly sprang from the spacecraft. He considered several alternatives, including paint flaking off the capsule and fluid from the thrusters. He determined that the most likely source was cooling vapor from the life-support system.[53] These droplets would have been visible only at sunrise, when sunlight struck them at an unusual angle.[54] Even after an explanation emerged, Glenn's reaction to the unexpected sighting remained a memorable part of his flight.

As Glenn crossed the Pacific in his first orbit, he tried eating to address concerns that swallowing might not be possible in a weightless environment. He consumed applesauce squeezed from a tube without any difficulty.[55] "It's all positive action. Your tongue forces it back in the throat and you swallow normally," he concluded.[56] One by one, he was erasing fears about living and functioning in a weightless environment.

At the end of the first orbit, he discovered a problem in the automatic system controlling the ship's attitude. The capsule's attitude would shift to the right, and then "it would overshoot to the left and then it would hunt and settle down again somewhere around zero [with the pilot appropriately situated facing away from the direction in which he was flying]. The spacecraft would then drift again to the right and do the same thing repeatedly,"[57] he explained later. Glenn worried about the hydrogen peroxide fuel being wasted. He decided he had to protect the fuel supply, so he took control. Throughout the rest of the mission, he guided the system directly with manual equipment or electronically through the fly-by-wire system, which connected hands-on manual control with the fuel nozzles that were part of an automated system.

Taking on this task consumed a significant amount of time that would have been spent on other observations and experiments. The necessity for him to abandon automatic controls confirmed one of Glenn's often-repeated arguments: that human pilots had special value. They could override and correct technological problems. A spacecraft control system works on three dimensional planes. Using the horizon as a guide, Glenn could easily control two of those three planes—pitch and roll. Yaw was a different matter. Imagine an airliner. Pitch is when the nose goes up or down. Roll is when one wing rises while the other goes down. Yaw is when the airliner's nose turns to one side or the other—right or left, east or west, north or south. "I found that the best way I could set up yaw was looking out the window and aligning with some object that was going straight away from me. The procedure that seemed to aid this more than anything was pitching down," he said.[58] His many years of piloting experience helped Glenn resolve this problem. Still, it was a struggle, and it would have been impossible for most humans to understand, much less calculate the proper attitude on all three planes.

Even with the attitude control problem, Glenn found time to make meteorological observations requested by the US Weather Bureau Meteorological Satellite Laboratory. He found it startling to see how much of Earth remained hidden beneath shrouds of clouds. On the dayside, he found it easy to identify different kinds of clouds, to follow storms, and to track rivers. In places, the cloud cover was thick, monolithic. During his hours in space, clouds covered massive areas, including East Africa and the Indian Ocean. At night, he had no trouble spotting lightning as it bounced from cloud to cloud or illuminated thunderheads from within. He believed that a rainbow of sea colors he saw in the Atlantic was a visible manifestation of the Gulf Stream coming out of the Gulf of Mexico. He also thought he saw the wake of a ship. Although the ship itself was not visible, he could see a V being carved into the ocean's waters.[59]

Without question, his favorite sights were the sunsets. He happily photographed a view that he knew would interest astronomers. He watched as the perfectly round sun seemed to flatten when it reached the horizon. At that moment, he said, "a black shadow of darkness moves across the earth until the whole surface, except for the bright band at the horizon, is dark."[60]

In one of his lighter moments late in the flight, Glenn asked Cooper to "send a message to the Director, to the Commandant, US Marine Corps, Washington. Tell him I have my four hours required flight time in for the month and request flight chit be established for me."[61] This chit was a sheet of paper that credited a flier with meeting his minimum monthly time piloting. For flying four hours in February, Glenn would receive an extra $245 in income.[62]

On Earth, however, an ominous mystery was unraveling and spreading as each tracking station picked up a disturbing signal showing that Glenn's heat shield and landing bag were no longer in the locked position. The heat shield was all that stood between Glenn and a fiery death as he reentered the planet's atmosphere. Although Kraft said his gut "told me instantly that it was a faulty signal,"[63] he anxiously awaited a recommendation on how to offer extra protection for Glenn. At one point, he reportedly yelled at a NASA technician, "Either you give me a decision, or I'm going to make one myself."[64]

Kraft and Shepard decided not to tell Glenn about the signal.[65] Glenn did not ask any questions on the second orbit when his radio contact in the Indian Ocean urged him to keep his landing bag in the off position, and he did not balk when subsequent capcoms repeated that message. He merely answered in the affirmative. It was only in his third orbit that the idea crystallized within him that something was wrong and that the men on the ground knew something he did not.

Kraft ultimately received a proposal to leave the retro-rockets attached to the heat shield instead of following standard practice and jettisoning them. The retro-rockets might hold the shield in place, but if any of the rockets held even a trace of remaining solid fuel, an explosion could rip apart the capsule. Death by explosion or death by fire were two possible futures for John Glenn. Kraft was wary but chose to retain the retro-rockets.[66] Mercury operations director Walt Williams and others swayed Kraft's decision in that direction. Williams later explained to the press that they used "the straps of the retro pack to hold the heat shield in place until we had enough aerodynamic force on so there would be no possibility of the heat shield dropping off."[67]

Colonel "Shorty" Powers, who served as the voice of NASA throughout the flight, told members of the media that the agency had detected "an indication of a problem with the heat shield deployment switch. The signal apparently was erroneous."[68] This was wishful thinking, and yet even this intentional miscommunication provided more information to civilians on the ground than Glenn had, and the announcement drew public attention to the impending dangers of reentry.

Glenn got the first direct evidence of a problem in his third orbit contact with fellow astronaut Wally Schirra in Point Arguello, California. Schirra helped Glenn with the timing of the retro-rocket firing that would guide him to splashdown in the recovery zone and told him that he should not jettison the retro package until he reached Texas. When the time came to fire the retro-rockets, Schirra asked Glenn if he was sure all three were firing; he said, "Are they ever! It feels like I'm going back toward Hawaii."[69] This, at least, reduced the danger of an explosion on reentry. After the retrofiring, Glenn again had to use precious fuel to reposition the capsule because it had been thrown out of place.[70]

The flight controller in Corpus Christi advised him that mission control recommended not jettisoning his retro package at all and started giving him instructions about steps he should take to achieve reentry with the retro package attached. At that point, Glenn calmly but impatiently asked, "What is the reason for this? Do you have any reason?" He was told that he would have to wait until he talked to Cape Canaveral for more details.[71]

Minutes later, as Glenn neared the Cape, Shepard spoke routinely by telling him to move into reentry attitude and then said, "While you're doing that, we are not sure whether or not your landing bag has deployed. We feel that it is possible to re-enter with the retro package on." Glenn absorbed this news without missing a beat. His only reply was, "Roger, understand."[72]

The message was clear now: mission control feared that his landing bag had deployed and that his heat shield was loose but might be held in place by the retro package. He reported that he was using fly-by-wire to maintain his reentry attitude. Then, his heartbeat steady, Glenn entered the reentry blackout zone, where no communication was possible.

As the ship initiated its fiery journey, he felt a thump and saw an orange glow outside. "Flaming pieces were breaking off and flying past the spacecraft window. At the time, these observations were of some concern to me because I was not sure what they were. . . . I thought these flaming pieces might be parts of the heat shield breaking off." The temperature outside was four thousand degrees. Still, he focused on preparing to manually complete the capsule's final slow roll. "The next 15 or 20 seconds ticked off like days of a calendar," he said.[73] He soon knew that what he had witnessed was the blazing destruction of the retro package.

When he emerged from the ionization zone that had disrupted communications, he said, "A real fireball outside!"[74] Relief washed over mission control, though the engineers knew that Glenn was not safe yet. The capsule was jolting back and forth. Glenn tried to stabilize it by using fuel that was running low. Both fuel tanks ran dry before the drogue chute deployed, and the ship started to oscillate severely. Just as Glenn was preparing to open the chute manually, it unfurled almost miraculously at an altitude that was seven thousand feet higher than expected. The oscillations ended, and Glenn received a reminder from the ground to deploy the landing bag.

Later, NASA released filmed images of Glenn recorded by a camera inside the capsule. He showed emotion only once—when he smiled at the sight of the main parachute blossoming above him.[75] The next day, he said, "That's probably the prettiest sight you ever saw in your life."[76] Pretty though the main chute might be, Glenn feared that he was descending ten feet per second too rapidly. He checked for rips or holes and discovered none.[77]

Below him, in the Atlantic, three aircraft carriers awaited. Positioned more than two hundred miles apart, each carrier had support ships to offer assistance. More than twenty other ships and fifty search-and-rescue craft were in place in the Atlantic in case the landing was hundreds of miles off target. Throughout the flight, additional aircraft had awaited orders to take off from airstrips in the Pacific in case Glenn's flight had to be cut short.[78]

Before reaching splashdown, he and the recovery force made contact via radio. Then the capsule hit the water with "a good solid bump."[79] It submerged and wavered from side to side before finally establishing its equilibrium atop the waves. He splashed down forty miles from the target zone, four hours, fifty-six minutes after liftoff, and initial explanations for missing the bull's-eye suggested that in establishing details of the retro-rocket firing, someone had failed to take into consideration the capsule's reduced weight, given fuel consumption during the flight. This mission was very much an opportunity for NASA to live and learn.

Soon, Glenn heard from the nearest recovery ship, the USS *Noa*, a destroyer. About six miles away, the crew of the *Noa* had seen *Friendship 7*'s descent and expected to arrive within minutes. Much more than at any time during his flight, Glenn felt uncomfortable swaying in the waves. The entire reentry had left him overheated, and after he hit the water, snorkels built into *Friendship 7*'s wall had opened, allowing fresh air to be pumped inside—in this case, warm, humid air. Sweat coated the overdressed astronaut. He had accessed the capsule's survival kit and considered escaping the stiflingly hot capsule, which felt like an overheated tin can. After some thought, he decided that exertion probably would only worsen his condition.

The *Noa* came alongside the capsule and a boatswain's mate attached a line to the capsule so that it could be lifted onto the deck within twenty minutes of splashdown. Once he was safely aboard, Glenn moved to begin the process of removing paneling so that he could exit through the top of the capsule; before long, he realized that disembarking that way would be too time consuming for a man overcome by heat. Instead, he told the ship's crew to stand clear so he could blow the hatch. He removed the hatch detonator, hit the plunger with the back of his hand, and blew it open. A cut on one finger from the plunger's recoil represented the only injury sustained in his mission.

Crew members pulled a smiling Glenn from the capsule onto the ship's deck. "It was hot in there," he reported.[80] He stepped out of the capsule at 2:20 PM. Powers reported from NASA, "John Glenn has left the *Friendship 7* spacecraft. He is reported to be a hale and hardy astronaut after having successfully flown the U.S. first manned orbital flight. U.S. Astronaut John Glenn is on the deck of the USS *Noa*."[81] His footprints were preserved in white paint on the *Noa*'s deck to record the historic moment.[82]

Glenn eagerly replaced his space suit with a lightweight flight suit and sneakers. Then he received a radio-telephone call from President Kennedy welcoming his return. One of the president's closest aides, Lawrence O'Brien, said, "The pride as president that finally we'd accomplished something in our quest to move up to the Russians, and his personal reaction was one of great excitement."[83] If JFK was excited, the response from lesser mortals went far beyond that.

Before Glenn met that near-hysteria, he had to receive the necessary medical approvals and dissect his flight with NASA. His already long day was far from over. Aboard the *Noa*, he got a quick medical check. Beyond complaining about the heat, he acknowledged not "stomach uneasiness" but "stomach awareness" in the minutes he had spent bobbing in the water. He said he had suffered no nausea, and the doctors concluded that a combination of heat, humidity, and dehydration had brought Glenn's flight to a less-than-perfect ending. Routine tests showed that he had lost more than five pounds since his early-morning physical and had sweated profusely after splashdown. His body temperature was 99.2, a degree higher than his first measurement of the day, but his blood pressure was only slightly higher than earlier readings.[84]

The captain of the *Noa* told Glenn he had been named an honorary member of the crew as well as sailor of the month. Glenn donated his prize money—fifteen dollars—to the ship's welfare fund.[85] He regretted that protocol required him to spend most of his time on the *Noa* alone, talking to a tape recorder rather than mingling with the sailors. The recording device served as the first of many NASA debriefers he would face.

About two hours after his arrival on the *Noa*, a helicopter hoisted him aboard for a quick flight to the USS *Randolph*, an aircraft carrier. There he underwent a chest X-ray and an electrocardiogram. He also began his technical debriefing.[86] Hours after his arrival on the *Randolph*, he climbed aboard a plane to Grand Turk Island. Entering the plane, the tired astronaut said, "I'm gonna sit back and relax and let someone else drive."[87] He would spend two days at Grand Turk as part of the debriefing process and extended medical examination. The doctors, who kept Glenn up until almost 1:00 AM, declared that there was "no sign of physiological change. John is in excellent condition."[88]

The next day's debriefing took half a day, and then he and Carpenter found time to go scuba diving, leisure time that was abbreviated because the still largely anonymous Carpenter saved another diver's life. Among the doctors checking Glenn's condition was psychiatrist George Ruff, who asked Glenn to fill out a form that he had completed many times after training exercises. The last question asked, "Was there any unusual activity during this period?" Glenn answered, "No, just a normal day in space."[89]

In the days, weeks, and months that followed Glenn's mission, NASA engineers sorted out some of the flight's troubling issues. They determined the problem that prompted the erroneous heat shield signal was a dented and bent "poorly machined shaft." Better quality control would have eliminated the problem.[90] A clogged yaw attitude control jet apparently caused the malfunction of the automatic attitude system.[91]

Glenn returned to Earth with a fair portion of angst. He had been shocked that NASA failed to inform him about the heat shield warning light. In his debriefings and other meetings, Glenn made his feelings known within NASA. As Carpenter observed, "Everyone saw John's eyes narrow as he appraised Kraft across NASA conference tables. The flight director was on notice, and everyone knew it."[92] Years after the flight, Kraft admitted, "We could have killed John Glenn just as surely as not."[93] It was that close. With all NASA's efforts to be prepared, it lacked sufficient redundancies to bring Glenn home safely without a lot of guesswork. Luck and smart engineering were on his side that day.[94]

Early on launch day, when Glenn had donned his spacesuit and rode the elevator to his space capsule, America had a great deal to learn about space travel. By the end of the day, his experience had answered many questions. He had confirmed the importance of human input when technology fails, and he had proved that human beings could function efficiently, creatively, and intelligently during periods of weightlessness. Both his ability to improvise in space and the creativity of mission control engineers after the heat shield warning signal fore-shadowed the near miracle of Apollo 13's safe return from its aborted lunar landing flight in 1970.

Just three years before Glenn's flight, some scientists had recom-mended that anesthesia or tranquilizers be given to humans to inca-pacitate them during space travel.[95] Glenn had proved the folly of that idea. Almost every action he took in space had been the stuff of science fiction until the moment he did it. In less than five hours, he had laid the groundwork for all the missions that would follow, including those that would take men to the moon.

6

MEANWHILE ON EARTH . . .

With their thoughts riveted on John Glenn, Americans came together on February 20, 1962. They listened, watched, dreamed, feared, and prayed. With the help of radio, television, and NASA communications, Glenn's flight created a network connecting Americans—East and West, North and South. Through national broadcasts, a college professor in Miami, a cab driver in San Francisco, a millionaire in New Orleans, and a farmer in the rural Midwest had equal access to what was happening in outer space, as long as they had access to a TV or a radio. All interested Americans experienced something precious, a day like no other, a day that would change America's mood and remake Glenn's life. With the help of relatively new gadgets, such as portable TVs and transistor radios, the singular historic event represented the most broadly shared experience in US history to date. The *New York Times'* James Reston spoke for many Americans, calling it "the greatest American ride since Paul

Revere . . . and it was pure American from start to finish: part Hollywood spectacular, part circus, part county fair—three times around the world in living color and news from heaven all the way."[1] And people all over the world who had access to the same media became vicarious travelers, following the calm and steady voice of John Glenn, speaking from so far away but sounding so close.

From the second his rocket lifted off the launchpad, Americans cheered Glenn. The sands of Cocoa Beach, Florida, were covered with fifty thousand screaming people. Many had children on their shoulders. They jumped, whistled, or performed their own victory dances.[2] With sand beneath his feet and hope in his heart, one man shouted, "Go, baby, go!" Without having to read the *New York Times'* analysis that it was "one of the most dangerous trips ever taken by a human,"[3] they could feel the danger as they watched the fiery rocket rise, and they celebrated the man brave enough to fly above sky-high.

After watching the liftoff, much of the crowd sat quietly, thoughtfully. "They watched the sky until there was nothing left to see except pelicans and seagulls," Gay Talese reported for the *Times*. As the minutes passed, they sat slumped over radios, listening for updates.[4] They had chosen to be on this beach, to see the rocket's fire, to hear the crowd as it thundered, to own a moment in history, but now, like other Americans and Glenn followers around the globe, they could do nothing to resolve growing suspense more heart-stopping than any story created by mystery maven Agatha Christie or director Alfred Hitchcock. What would happen next was a story that would unfold minute by minute. There had been no preview of this never-before-seen event, no opportunity to measure the odds. As one adventurer did his job, much of humanity followed his progress.

"Breathlessly, Americans waited. With mounting tension, they clung to their television and radio sets" in a nation caught between terror and elation, the Associated Press reported.[5] Ralph McGill, editor and publisher of the *Atlanta Constitution*, was entranced by the rhythmic sound of Glenn's heartbeat heard during some of the coverage. Glenn was distant, too distant to see, and yet McGill could hear the steady pounding. He "realized that all over America in offices, in factories, in farmhouses and city flats there were hearts beating with this sound. All of our hearts contracted and expanded with that of John Glenn's."[6] Saul Pett, an Associated Press commentator, said that at 9:47 AM when Glenn was set to blast off, "We pushed with our bodies and we pushed with our minds and we wanted in the worst way for John Glenn to go in the best way. Go, go, go, go up John Glenn, go straight, go true, go safe."[7] A woman from Cicero, Illinois, told the *Chicago Tribune*, "My heart has been pounding as fast as it could. I felt as though he was my own son."[8] One commuter in Grand Central Station told a reporter, "We're all up there with him."[9] Telephone and man-on-the-street polls showed that residents of San Bernardino, California, felt a "personal attachment" to Glenn, and many stayed up overnight to watch coverage of events leading up to the liftoff three time zones away.[10] Some Americans truly lived each moment with Glenn, and the anxiety of that day was a communal experience—as the jubilation later would be.

Millions of Americans could not turn their backs on John Glenn. A radio or a TV was a constant companion for those who had one on this nerve-racking day. A Kansas City pharmacist kept his transistor radio playing all day. "I felt a little tension at blastoff time," Charles E. Byrne admitted.[11] A taxicab company in Charlotte, North Carolina, had a problem: dispatchers could not reach many of their drivers, who apparently were tuned in to the flight. When questioned about how Glenn's flight had affected business, an executive at Omaha's largest department store had a ready answer: "I don't believe there has been any business." Long-distance phone calls dropped precipitously in Portland, Maine, and a similar reduction in both local and long-distance calling was noted elsewhere across the country. In a Louisville courtroom, official business ceased when a defendant brought in a portable TV that everyone viewed. Even Reno bettors walked away from the tables to watch another man's gamble.[12]

When John Wanamaker's department store opened at 9:30 AM in Philadelphia, about one hundred people dashed into the TV area, where fifty-six sets were on display. By liftoff, the crowd had quadrupled. Even an hour after Glenn's splashdown, about one hundred viewers remained in the store watching postflight updates. Philadel-phia bars were far busier at midday than coffee shops for one simple reason: most bars had TVs, and most coffee shops did not.[13] Some Northeast Philadelphians spent the first few minutes of the flight in confusion; a Philadelphia Electric Company crew needed to make repairs, and much to the dismay of the residents, the workers chose the moment twenty-five seconds before liftoff to shut down power to thirty homes. By the time electrical power returned, Glenn was flying high above Africa.[14]

In Atlanta, where many people carried portable televisions to work, sidewalks were less congested than usual, and a marine recruiter proudly proclaimed, "I've always contended the Marines were out of this world."[15] The flight triggered an entrepreneurial impulse in the proprietor of Sacramento's Clip Top barber shop. Tom Harris posted a sign, GET JOHN GLENN HAIRCUTS HERE, and afterward, he reported that business had been brisk. "It's a close haircut. It's not a flat-top and it's not a butch, but in between the two."[16] A girl from Shreveport, Louisiana, admitted to a reporter, "My mother asked me this morning if I'd like to go up there or to the moon or somewhere and I told her I'd be scared to death and I would, but it was wonderful and awfully exciting for Col. Glenn to do it."[17] In Nashville, a mother of four said, with some foreboding, "I can see all the children want-ing to eat their food out of toothpaste tubes," in the way Glenn had consumed applesauce.[18] And at Saint Peter's Church in Saint Louis, a women's group heard from Robert F. Hage, a vice president of McDonnell Aircraft, the company that built the *Friendship 7* capsule. Hage assured his audience that he had searched the Bible and found "no evidence or warning or moral reasons for restricting ourselves to earthly exploration."[19]

Even crime seemed to be taking a holiday in Chicago. The emer-gency switchboard at police headquarters was almost silent. The number of empty seats on buses and trains was higher than usual. A Chicago cabbie with no radio in his car, rolled down his window and yelled, "How's he doing? How's our boy?" Primly dressed suburban housewives found themselves at midday crowded around the bar in

Thousands stand before a gigantic TV screen in Grand Central Station as Glenn takes off. *Getty Images*

the Iron Horse cocktail lounge to watch the TV. A pharmacist stepped from behind the pharmacy counter to declare, "He's over Texas." And a Polish janitor, once a refugee, took a break from hauling a steel trash can and said, "Boy, he makes it. We show those Russians now!"[20]

In New York City's Grand Central Station, Midtown workers and weekday shoppers stopped in their tracks and stood frozen, staring at a huge twenty-five-foot square TV screen provided by CBS News. Late for work? Maybe, but many could not leave. They were enthralled. A shoulder-to-shoulder crowd that sometimes reached seven thousand stood transfixed. Eyeballs glued to the screen, they squirmed and shifted nervously as they felt the minutes ticking onward. At the moment of liftoff, more and more people angled their bodies into the area in front of the gigantic monitor. TV newsman Douglas Edwards reported that despite the mass of humanity, you could have heard a pin drop during the final countdown. Then came screams, applause, hoots. CBS called it the largest stationary crowd ever in Grand Central.[21] At liftoff, some cried, "Go, go, go!" Within the dense crush of Americans, some people's eyes overflowed with tears.[22]

Eventually, some people left, of course, to report to work, and others took their places. Some chose to spend their lunch hours in the station standing before the big screen. Commuters on subway platforms and aboard trains heard the news via loudspeakers. When Glenn splashed down and contacted the recovery vessel, the Grand Central crowd applauded and cheered in happiness—and relief. The *New York Mirror* reported that in the station, "before and during the stab into space, it was 'God be with you' and hope. After, it was 'Thank God' and elation."[23] As soon as Glenn's recovery was assured, officials at Grand Central asked CBS to turn off the TV so that evening commuters could get to their trains more easily.[24]

Around lunchtime in Manhattan, John Glenn was in his second orbit, and those who wanted to follow his flight without joining the orderly mob scene at Grand Central filled bars around Times Square. Others stopped to buy a cup of coffee, hoping to get a quick update before returning to work. On New York–area college campuses, transistor radios were the must-have accessory for students on the go. On sidewalks, normally anonymous, brusque, and fast-moving walkers slowed as strangers reassured one another: "He's safe" or "I just heard he's OK" or "He's aboard the destroyer." At about that same time, Walter Cronkite quoted sources on Madison Avenue as saying, "There's more confetti there than has been seen on a Midtown street before." Later, he reported, "The whole nation seems to be going wild in celebration."[25] Workers with offices in Rockefeller Center tossed ticker tape and pages torn from phone books into the wind, and the debris floated for several blocks across the city.[26] Cascading tape from adding machines and other paper scraps flowed around the green-glassed Secretariat Building at the United Nations. Office workers—not UN representatives—took the blame for this breach in protocol. Representatives from Somalia and Turkey crowded in the Press Club to watch the latest news with US ambassador Adlai E. Stevenson.[27] On the floor of the New York Stock Exchange, the news of Glenn's recovery arrived after 3:00 PM. Slowly, mumbling conversations grew into a roar of exuberance, but trading barely paused.[28]

Many Los Angeles residents followed the flight, with somewhat hazardous results: when Glenn lifted off, there was a flurry of small traffic accidents on both city streets and freeways. Some drivers yelled out their car windows, while others honked their horns. When the USS *Noa* lowered Glenn's capsule to the destroyer's deck, church bells tolled.[29] One L.A. woman was in such a hurry to get to her TV when she awoke on this morning that she fell and fractured her right leg.[30] The *Los Angeles Times* reported that Glenn's flight was "undoubtedly the best-watched TV show in local history." News-hungry consumers quickly emptied newsstands selling midday editions of the *Times* with a banner headline that said simply, GLENN GOES.[31] In Los Angeles' Police Building, officers crowded around a television in the press room. Captain Joseph E. Stephens told a reporter, "We saw it all from beginning to end. We were an eyewitness to history."[32] In Northern California, even death row prisoners at San Quentin watched when their recreation periods overlapped with Glenn's flight.[33]

Lewis Fussell, science director for defense contractor EG&G, was ready to take part in his home city of Las Vegas's favorite pastime: betting on future outcomes. He said, "In ten years, the odds are very good we will have rockets at least going around the moon and perhaps in interplanetary space." Colonel Joe B. Russell, marine barracks commanding officer at Lake Mead Naval Base, admitted that he had carried a transistor radio to work so that he could follow the flight. "We're glad to see the delays are over and I know they were quite trying for everyone involved," he said.[34] Clayton P. Shepard, a minister in Battle Creek, Michigan, composed a poem about Glenn and mailed a copy to Lyndon Johnson.[35]

Annie Glenn and her children, Dave and Lyn, stayed at home in Arlington, Virginia, huddled in front of the TV set. Annie's parents and family friends joined them. Dave had several documents provided by his dad to help him follow the flight.

After Glenn's recovery by the *Noa*, his wife and children came out of the house and spoke to reporters from their front porch. Annie, who specialized in short comments, described herself as "very happy."[36] She told reporters that her worst moments probably occurred in the last fifteen seconds of the countdown. Her children were proud of their father's achievement, she said.

Glenn called his wife and children around 4:40 PM. Soon after that, President Kennedy called to congratulate the whole family. In the evening, Annie, Dave, and Lyn attended a celebration at the home of Glenn's World War II buddy and next-door neighbor, Lieutenant Colonel (and future General) Tom Miller.[37] Glenn's parents watched the liftoff at home in New Concord, Ohio, and then joined an excited crowd at Muskingum College.[38]

Nationwide, children and their teenage brothers and sisters followed the flight in school. In Los Angeles, thousands of youths watched and heard the flight. Many rose before the sun to start tracking developments at home. L.A. school administrators reported, "Everyone is keyed up," especially science instructors. One school, Sherman Oaks School, had seven televisions so that children could follow *Friendship 7*. Teachers used bulletin board exhibits to track Glenn's flight for students at Calabash Street School in Woodland Hills. Sixty-five Calabash fourth and fifth graders had written to the astronaut in the previous weeks to give him their best wishes and to ask that he write to them about his mission.[39] Saint Vibiana's Cathedral School students prayed at 8:00 AM mass for Glenn's survival and success—and other churches, too, said special prayers for the astronaut.[40]

At Western High School in Las Vegas, students caught snatches of news coverage between classes from the school's loudspeaker system, which was transmitting radio reports.[41] Sixth-grade students at Kalihi Waena School in Honolulu expressed so much anxiety that their teacher attempted to focus their attention on where Glenn was in space, by using a globe to plot his course. And for these students, whose home had been one of the nation's states for only a bit more than two years, there was pride in knowing that the Kauai tracking station assisted Glenn in flight.[42]

Cincinnati principals received no mandate from the school system about how to handle this special day: some provided radio coverage via public address systems; others turned on TVs; a few schools even allowed students to listen to their transistor radios through ear plugs.[43] More than five hundred thousand Chicago students watched the lift-off on TVs in cafeterias and auditoriums. "It was like being a witness at the dawn of a new era," said William S. Eaton, principal of Darwin Elementary School.[44]

At one Pennsylvania school, a small group of students got permission to leave their classes and watch the flight. A high school senior described the scene:

> The countdown, nearing its last few minutes, produced an extreme hush. A short hold caused a spontaneous moan of disappointment. After resuming, the final seconds slipped by in a painful quiet. Many of us found it difficult to control the tears and earnest prayers we held for [Glenn], and more selfishly, for our future also. During the first minutes of flight, the utter silence reflected our hopes, and our thoughts of [Glenn] followed us all during the day.[45]

Ten-year-old Mark King in Mount Vernon, Ohio, kept his fingers crossed throughout the entire flight.[46] Fourth grader William Maiben of Salt Lake City worked all day so that he could "put his paper to bed" with complete details of Glenn's flight, and that edition was distributed to fellow students at Morningside Elementary the next day.[47]

One of the nation's oldest institutions—the Post Office Department—tied its hopes to the dream, and for the first time in history, it released a new stamp on the day of a big event without forewarning the public. New stamps commemorating the flight went on sale after Glenn's splashdown. In Atlanta, they became available just after 4:00 PM. Unfortunately, a rumor that stamps would be sold all night at one post office sent hundreds of Glenn fans and stamp collectors on a

President John F. Kennedy, Vice President Lyndon B. Johnson, and lawmakers watch Glenn's launch. From left to right: Representative Hale Boggs of Louisiana, Speaker of the House John W. McCormack of Massachusetts, Representative Carl Albert of Oklahoma, Senator Hubert Humphrey of Minnesota, Johnson, Kennedy, and Senator Mike Mansfield of Montana. *John F. Kennedy Library and Museum, Boston*

wasted journey.[48] Some cities, such as Jefferson City, Missouri, put the stamps on sale as soon as Glenn's flight was officially declared completed—at 2:45 PM. The four-cent "U.S. Man in Space" stamp showed a capsule with Earth below.[49] James Webb, NASA administrator, justified the issuance of the stamp, writing, "This achievement symbolizes the determination of the United States to lead in space—to explore and develop its wonders for the benefit of all mankind."[50]

Government officials kept track of Glenn's flight in various ways. President Kennedy, who had met with Glenn privately and liked him, arose early to watch coverage of the liftoff. He turned on his TV at 7:15 AM, watching without sound while eating breakfast and making his morning sweep through this day's newspapers. JFK, who had planned to talk to Glenn during his flight, changed his mind after learning that the astronaut was wrestling with significant problems. The president carried on his daily business, though he often peeked at a TV.[51] When he spoke to Glenn following his safe recovery, he said, "We are really proud of you."[52] In a public statement afterward, he told the American public, "The impact of Colonel Glenn's magnificent achievement goes far beyond our own time and our own country." Speaking of space, he said, "This is the new ocean, and I believe the United States must sail on it."[53] (Despite his excitement about Glenn's flight, Kennedy had other things on his mind: He was bothered by what appeared to be a posed photograph of himself, the president of the United States, on the cover of *Gentlemen's Quarterly*. The slick magazine showcased the latest men's fashions—and many considered modeling clothes to be appropriate for women, but not for "real men." On a day when many saw John Glenn as a warrior proving his manliness, the president appeared to be a fashion model.)[54]

Secretary of Labor Arthur J. Goldberg postponed a speech to be delivered to union leaders, who made no complaint about gaining free time to watch flight coverage. Secretary of the Treasury Douglas Dillon and Secretary of Commerce Luther Hodges unashamedly spent the day with portable TVs nearby, and Secretary of the Interior Stewart Udall began a 2:00 PM press conference with a plea to end before the scheduled time of Glenn's splashdown.[55] On a day that began with a White House breakfast and ended with a white-tie dinner at the same place,[56] Vice President Johnson took time to write a letter to Glenn's son, Dave. He told the sixteen-year-old, "It is a day as important to the world as the day 470 years ago when Christopher Columbus set sail for unknown horizons." He called Glenn's flight "the greatest 'voyage' our nation has ever undertaken."[57] On this afternoon, the US House twice halted debate on the debt limit to cheer Glenn's return. Faced with a nearly empty chamber, the Senate's leadership had adjourned at 2:30 PM.

After splashdown, many cities and towns spontaneously issued proclamations, and some renamed streets in Glenn's honor. In Trenton, New Jersey, an impromptu parade honored Glenn. Army, air force,

Mayor J. Randolph Manahan dedicates John Glenn Drive during May 1962 in Morristown, New Jersey, with a marine honor guard.
US Marine Corps

and high school bands added music to the celebration, and Governor Richard J. Hughes declared "John Glenn Day."[58]

Meanwhile, the US State Department found a way to use Glenn's flight to cover up the use of reconnaissance satellites. The United Nations had a regulation that required nations to report incidents of "sustained flight." Glenn's flight lasted relatively few hours, so the United States opted not to register it, trying to set a precedent for not registering future flights with comparable durations—namely, those of recon satellites. (The ploy fooled no one.)[59] The Soviet news service TASS reported in late March that the United Nations reacted with "surprise and suspicion" to the US decision not to report details of Glenn's flight. TASS referenced a *New York Times* report that the United States wanted to divert publicity from its use of "spy satellites."[60]

At Cape Canaveral, Representative James G. Fulton, a high-ranking Republican member of the House Space Committee, criticized

NASA for this day's delays in Glenn's flight. He said, "The time has come when the common garden variety of failures in missile countdowns must be over." Fulton, a Pennsylvanian, said that the day's postponements represented issues that were "the equivalent of plumbing."[61] True, one flaw was a broken bolt in the hatch, which could seem like a minor problem to many people. However, to the man inside the capsule, any flaw in the equipment was potentially a matter of life or death, and this day's delays included more complex technological problems involving rocket guidance and the loss of power in the Bermuda station, which was a key location in the worldwide network of communications bases maintaining almost constant contact with Glenn.[62]

In other news on that day, Secretary of Defense Robert McNamara reported that US military aid to South Vietnamese troops was "becoming more and more effective." In later years, after Kennedy's successor, Johnson, sent half a million troops to South Vietnam, McNamara eventually concluded that US involvement in Vietnam had been a wrong-headed policy, but on this day, he was "delighted with the progress" made over recent weeks.[63] President Kennedy asked Congress to raise the pay of white-collar federal government employees. Over three years, his plan would amount to a $1 billion pay raise.[64] JFK lost an important test in the Senate when his administration tried to force quick Senate action on creation of a cabinet-level department of urban affairs and housing. The push for hurried consideration lost by a 58–42 vote.[65] (Kennedy had promised to name an African American to lead this department, and southern and western Democrats had joined conservative Republicans to block its establishment. The Department of Housing and Urban Development would not be approved until 1965, during the Johnson administration.)

People in many parts of the world followed Glenn's flight. The Voice of America made that easy, beginning uninterrupted coverage in English at 7:00 AM, almost three hours before Glenn's liftoff. The agency broadcast in thirty-five other languages, eleven of which were native to the Eastern European Soviet bloc. As VOA employees were sharing the flight with the world live, the agency also hurried to send out a one-reel film of Glenn's trip in forty-one languages to

107 nations.[66] Millions of western Europeans followed the flight from beginning to end. Even in Moscow, reporters broke in to regular programming to offer updates, and officials at the Voice of America noted that the Soviet Union chose not to jam American reports aimed at Iron Curtain nations.[67]

The flight won a banner headline and consumed about two-thirds of the *Sydney Morning Herald*'s front page, which could report only that Glenn had finished his first orbit, had seen the lights of Perth, and had crossed Australia in just nine minutes.[68] Thousands of Aussies delayed their bedtimes so that they could hear the launch broadcast from Cape Canaveral via Voice of America. They heard calm news commentators throughout the countdown and its delays, but when the rocket took off, listeners were startled by shouts: "It's beautiful, it's a beautiful launch!"[69]

Among those who listened were several Yugoslavian students in the town of Zrenjanin. They stayed by the radio from start to finish.[70] A young Mexican man named Pedro Orozco, who had received his bachelor of science degree from the University of Arizona, excitedly followed Glenn's progress by radio.[71] Truong Thanh Binh, a youth in South Vietnam, had heard reports about preparations for Glenn's flight as well as the flight itself. He wrote Glenn, congratulating him on his unflappable behavior. He told the astronaut that people all over the globe had stopped to listen to news of the flight and had rejoiced over his success.[72]

Americans reveled in Glenn's triumph, and they found something to celebrate in the federal government's openness about the details of the flight. "Col. Glenn's flight today is something in which every American can take pride," Los Angeles mayor Sam Yorty said. "It makes us even prouder to realize that the American space effort was carried out publicly." City Supervisor Frank G. Bonelli agreed, saying, "We should mark the great contrast between the American method of open and objective reporting of every step in Glenn's flight compared to secrecy maintained by the Russians."[73] The Vatican's daily newspaper, *L'Osservatore Romano*, called the American approach "the blaze of public participation."[74]

A *New York Times* editorial called the flight "one of our finest hours" and said, "In a sense, time stood still as countless millions watched and indeed participated in one of the greatest dramatic events of modern times."[75] Michigan's *Battle Creek Enquirer* compared Glenn to an earlier hero: "Like another daring American, Charles Lindbergh, who blazed a new trail for other men to follow 35 years earlier, John Glenn has carved for himself a permanent niche in history."[76] Atlanta's Ralph McGill called February 20 a "therapeutic day," allowing Americans to put aside pettiness and greed as they became swept away by a rare opportunity to experience "greatness of spirit, and the nobility of man at his best." He noted that Americans, in that moment, viewed Glenn as someone traveling between America's present and its future. "We thought of him as flying through the limitless skies of night into a tomorrow which had not yet come to us. We welcomed him back into our today."[77]

All across the country on this day, Americans witnessed the usually businesslike Walter Cronkite becoming a bit of a mess over almost ten hours of continuous coverage from an elevated makeshift newsroom about two miles from the launchpad. For Cronkite, sitting inside the glassed-in newsroom/studio under the Florida sun was too much. During a commercial break, he removed both his jacket and tie. When the audience returned, the sweaty anchorman apologized for his attire, explaining that he was baking in an oven. Clearly, this kind of extended live broadcasting was less polished than it would become in later years. Several times, the cameras went live more than a few seconds before Cronkite started talking. Close to the end of Cronkite's coverage, a camera apparently fell from the elevated newsroom to the ground below. A visibly startled Cronkite reported that CBS had just lost a camera. Network staffers in an adjoining glassed-in room seemed oblivious to the cameras still operating as they rushed to a window to stare down at the ground.[78]

All three major networks followed Glenn's day from dawn to dusk—and beyond, in some cases. For much of the day, ABC, CBS, and NBC pooled their resources so that all had access to more cameras. For TV viewers, the coverage had a bewitching effect, as time seemed to move fluidly backward and forward. Listeners and viewers would be told that Glenn was flying over one continent, and then NASA would release a five-minute tape of his voice communications when and where he had been ten minutes earlier. Journalists had been told about some apparently resolved issues with the heat shield, and reports of what they knew raised tensions among the public during Glenn's descent. Nevertheless, the depth of the danger did not become clear to typical Americans until after Glenn had safely returned to Earth. His coolness under the threat of imminent death contributed to his new and expanding legend.

On that night, the evening news programs, which lasted only fifteen minutes in 1962, focused almost entirely on Glenn's story. CBS News, anchored by Douglas Edwards, found room for only one non-Glenn item, and both CBS and NBC bumped entertainment programming later in the evening to accommodate more news about Glenn.[79]

The Nielsen organization's official TV ratings for the flight's coverage did not emerge until a month afterward. The ratings numbers showed that 39.9 million homes watched TV reports on the flight, with the average home viewing about five hours and fifteen minutes out of approximately ten hours of live reporting. Thus, average household viewing exceeded the length of Glenn's flight. Not surprisingly, viewership peaked at liftoff and splashdown. Even during a midflight lull in viewership, this coverage drew five million more viewers than the top-rated entertainment program of the 1961–62 season, *Wagon Train*, typically attracted for a prime-time weekly episode. Glenn drew higher viewership than either the Democratic or the Republican National Convention in 1960. About the same number of people viewed flight coverage as tuned in to the 1960 presidential debates between Kennedy and Republican Richard Nixon, and the debates had aired in prime time, not during the workday.[80] Radio and television broadcasts of Glenn's flight had an estimated cost of $2 million.[81]

After a day of watching and analyzing, Rick Du Brow of United Press International concluded, "Unless you are made of stone, the TV coverage often became a deeply emotional event for the viewer. After weeks of frustrating delay, this steel-nerved 40-year-old father was ready to risk his life." He added candidly that on a day of such capti-vating history-making news, "the interesting thing about the TV cov-erage is that it really didn't matter how good it was."[82] Cynthia Lowry of the Associated Press declared, "It was a great day for television and one of which it can be proud."[83] Oregon's *Eugene Guard* wrote, "Prob-ably never before in all the world have so many people been so close to a minute-by-minute account of a great event in history."[84] Over the coming years, TV would demonstrate its ability to unify the nation, most often in times of tragedy. Glenn's flight generated a less common communal celebration.[85]

During the hours when Glenn flew high above Earth, he had no idea how many stowaways spent their hours traveling through space with him as they watched or listened to broadcasts. Hundreds of thou-sands had heard his heartbeat,[86] and now, as he would soon discover, they believed that he belonged to them. He was their hero.

7

SUPERSTAR

J ohn Glenn returned to Earth as something more than a marine lieutenant colonel: he wore the aura of an American star. The change in his life began to become clear as soon as he again stood on American soil. The entire nation—from pastors to schoolchildren, from congressional leaders to doo-wop-loving teens—seemed to fall victim to Glenn fever. The space sensation captivated Americans. To the young, he was a daring explorer. To adult Americans, he was more: they needed him, or someone like him, to boost the nation's morale. He was the right man for a country ready to kick up its heels.

The United States' last big victory extravaganza had come as the nation marched out of World War II with reasons to feel exultant. The nation's potential seemed boundless. Then, after the start of the Cold War in the late 1940s, the Soviet Union's postwar technological leaps in nuclear weaponry and space exploration began to cast a shadow over Americans' rosy picture of the world. Suddenly, totalitarianism looked almost as efficient as it was terrifying. Americans were dangerously close to developing an inferiority complex—and lionizing a hero was the perfect cure.

Though President Kennedy enjoyed the highest job approval ratings of any president since polling began in the 1940s, he and his brother, Attorney General Robert F. Kennedy, had become polarizing figures. Even First Lady Jacqueline Kennedy, who was beautiful, well dressed, and well educated, had her share of haters. Consequently, while the youthful, handsome, and polished JFK was a hero to many Americans, he was an irritant to some and a devil to others. Divisive and unavoidable issues such as civil rights made it difficult to please everyone. Thus Kennedy had little chance of becoming a unifying figure in a particularly divisive decade of change.

On February 19, 1962, Americans had no idea how to escape their funk. And then John Glenn accomplished something glorious; by achieving his mission and bravely facing possible destruction, he made Americans feel proud, honorable, invincible, united. The handsome man who loved his wife, worshipped his God, and served his nation seemed perfect. Never mind that the Soviet Union's second cosmonaut, Gherman Titov, had circled Earth seventeen times, and Glenn had orbited only three times. Glenn's mission changed everything, like a mood-enhancing drug for an entire nation. Suddenly even the most anxious Cold Warriors had reason to party.

The first of many roaring crowds greeted Glenn when he arrived from Grand Turk Island with Vice President Johnson on February 23 at Patrick Air Force Base in Florida. Glenn, still oblivious to the spectacle he had become, rushed off the plane to hug Annie, Dave, Lyn, and his parents. He turned from the crowd briefly to hide the tears he shed when reunited with his family. That display of emotion simply elevated national euphoria over this man—a courageous Cold Warrior who wept upon seeing his loved ones. He later wrote, "I couldn't hold back tears. Annie, most of all, had been at the back of my mind the entire flight."[1]

Johnson gave Glenn a taste of what was to come. "It's a great pleasure to welcome home a great pioneer of history," he proclaimed to a boisterous crowd.[2] Then Glenn's official party set off in a line of cars for the short ride to Cape Canaveral. The motorcade quickly became more like a parade. Standing along the route were approximately one hundred thousand spectators, including tourists, children, and moteliers.[3] Some Glenn fans threw confetti; others held signs proclaiming, WE ARE PROUD, JOHN and COLONEL GLENN, OUR NEW COLUMBUS. At one point, the photographers' truck almost collided with the car carrying Glenn, and he shouted good-naturedly, "I don't mind going into orbit, but I wouldn't want to ride in one of those trucks with you guys."[4]

Once the astronaut and his entourage reached Cape Canaveral, President Kennedy and members of the House and Senate space committees greeted him. Kennedy wanted to see the launch facility (which all-too-soon would carry his name after his death), and Glenn provided a tour for the president and vice president. He introduced Flight Director Chris Kraft and other NASA officials who had played key roles in his flight. Glenn quickly learned that Kennedy was more interested in the "sensations" of space travel than he was in the technical details.[5] Glenn and his colleagues gave Kennedy a missile hat, which he took with him. (JFK was notoriously opposed to wearing hats in public if there was even the slightest chance that they might make him look silly.)

In the Cape Canaveral crowd were many of the behind-the-scenes workers who had made Glenn's success possible. He shared the acclaim for the flight's achievements with all of them, calling himself a "figurehead."[6] At times, the crowd surged forward, threatening to topple Glenn, Kennedy, and Johnson. Alarmed, the vice president asked, "Who's leading us here?" Kennedy evaluated the situation himself and said, "All right, we'll wait right here."[7] Like children dodging the person who was "it" in a game of tag, avoidance became the three men's modus operandi. During one escape attempt, as Kennedy and Glenn headed in one direction, the crowd blocked their route, so they did an about-face to where they had begun. As they were scurrying for safety, a band started playing the "Marine's Hymn." Kennedy asked Glenn what loyal marines did when they heard the song, and Glenn told him that marines typically stood at attention. Instantly, Kennedy

stopped and stood ramrod straight. Glenn joined him.[8] The chaos at
Cape Canaveral caused the astronaut's father, John H. Glenn Sr., to be
separated from his family for nearly an hour as he fought to find his
way through the noisy and rambunctious mob.[9]

Addressing the gathering, Kennedy commented that the astronaut
spent this day being "launched into public orbit," and the president,
who was known for his wit, said Glenn was learning that "the hazards
of spaceflight only begin when the trip is over."[10] Kennedy honored
him with the National Aeronautics and Space Administration's Distinguished Service Medal, telling the crowd, "His performance was
marked by his great professional knowledge, his skill as a test pilot,
his unflinching courage, and his extraordinary ability to perform most
difficult tasks under conditions of great physical danger."[11] Kennedy
also predicted that Glenn's achievement would be long remembered:
"All of us remember a few dates in this century, and those of us who
were very young remember Col. [Charles] Lindbergh's flight, and
Pearl Harbor, and the end of the war—and we remember the flight
of Alan Shepard and Major Grissom, and we remember the flight of
Col. Glenn."[12]

Later in the day, Glenn addressed reporters at a lengthy press conference. The transcript of that session filled more than a full page of the
Washington Post. Faced with a room full of journalists, Glenn showed
none of the shyness displayed by other astronauts and delivered none
of the terse, jargon-filled answers that were common among his colleagues. He described many facets of his flight, telling the journalists,
"It was quite a day, I don't know what you can say about a day when
you see four beautiful sunsets in one day."[13] (He viewed three sunsets
from orbit and one after splashdown.)

Under questioning, Glenn admitted that he had been somewhat
alarmed after he learned that the heat shield, which was intended
to protect him during reentry, might be loose. He said, "There were
moments when I thought the heat-shield was breaking up," and in a
great understatement about the prospect of his own death, he added,
"That would have been a bad day all around." Glenn also spoke with
genuine wonder about the "fireflies" that he saw at first light on each

orbit. Trying to provide a full picture of the entire flight, he told reporters about the experience of eating applesauce from a tube, and said crumbly foods were a definite no-no in a weightless environment. He reported that he had no great sensation of speed and compared the experience to looking out an airliner window when the jet was cruising at thirty thousand feet.[14]

As soon as Glenn's workday ended, he and his family took the opportunity to relax after what had been a long Friday. They headed to Key West, where they spent a few quiet days together without the company of curious onlookers. For Glenn, there soon would be nowhere to hide from zealous fans. His smile and the auburn fringe surrounding his balding pate would become so well known to Americans and others around the world that, with reluctance, he had to accept no longer enjoying the privacy and anonymity shared by most Americans. He was now an autograph-signing idol.

While the Glenns were enjoying their respite before the storm, the Reverend Lloyd George Schell, a minister in Hornell, New York, delivered a sermon—not about Jesus or one of the prophets but about John Glenn. "What man could stand the tension of climbing into a space capsule and waiting over five long hours, before learning that the flight was off, and then go back to more intense training and indefinite waiting?" Schell asked, as he spoke about Glenn's faith and the calm voice that had communicated with mission control from orbit, a voice that "struck [Schell] with wonder." He quoted Glenn's religiously oriented response to the idea of life on other planets: "Conditions capable of sustaining life exist elsewhere. Seems to me that if God was capable of creating intelligent life on this planet, He could have done it on others. And I know of nothing in the Bible that says we are the only race of beings with minds and free will that He has created, or that Earth is the only place he's done it."[15] Not John the apostle, John the astronaut.

At the end of the weekend, the Glenns joined the Kennedys for a trip to Washington on Air Force One. Jacqueline Kennedy and her four-year-old daughter Caroline welcomed them. Caroline had heard enough about spaceflight to know that the first travelers on US rockets had been members of the primate family, so when her mother

President John F. Kennedy, right, escorts Glenn to the Oval Office for a reception in his honor on February 26, 1962. *John F. Kennedy Library and Museum, Boston*

introduced her to Glenn, she looked at him curiously and innocently asked, "Where's the monkey?" Glenn later told Congress that question held his ego in check.[16]

In Washington, NASA officials as well as the astronauts and their families (except Gordon Cooper, who had not yet returned from his communications post in Australia) attended a reception at the White House. Then the astronauts climbed onto the backs of convertibles and slowly rode to the US Capitol in a chilly rain. Despite the weather, thousands of soggy Washingtonians jostled for an opportunity to see Glenn and the other spacemen, who huddled in their jackets and waved at the damp crowd.[17]

Next up: a speech before a joint meeting of Congress.[18] As Glenn entered the chamber, lawmakers leaped to their feet with a shout. Although he had never graduated from college, Glenn mesmerized the nation's lawmakers. All smiles, leaders jockeyed for the chance to shake his hand, slap his back, applaud his words, and otherwise show respect for this man of wonders named John Glenn. The US Supreme Court had chosen not to attend President Kennedy's recent State of the Union address; however, the justices made sure to be in the Capitol when Glenn was there. *Life* reported, "Mr. Justice William O. Douglas sat like a youngster watching Buck Rogers, the sensations and emotions of spaceflight rippling across his face. Senator John McClellan removed his glasses to wipe his eyes."[19]

Glenn wowed the crowd with his humility, saying that thousands of hardworking and dedicated people deserved a share of the praise for the success of *Friendship 7*. He insisted that his predecessors in space, Alan Shepard and Gus Grissom, receive a round of applause; yet all eyes were on him. He explained to Congress that as wonderful as they thought his flight was, it was just a beginning. To clarify his meaning, he recounted an anecdote about British prime minister Benjamin Disraeli looking at the electromagnetic experiments of Michael Faraday and asking, "But of what possible use is it?" Faraday answered, "Mister Prime Minister, what good is a baby?" That was where the space program stood at that moment, Glenn told his audience—in its infancy. Amplifying that point, he added, "*Friendship 7* is just a beginning, a successful experiment."[20] He talked unashamedly about his love of God, family, and country. He demonstrated self-effacing humor. The lawmakers, cabinet members, justices, and the joint chiefs seemed to glow and jump almost as if Glenn's personal magnetism had given them a jolt. They stomped, they hooted, they whistled, and some shed more than a single tear.

It turned out to be another extremely busy day for the hero's family. Hours later, when the solitude-seeking Glenns reached their Arlington home, it was surrounded by about 750 well-wishers.[21] The Glenns and their children politely acknowledged the astronaut's new admirers, who were awestruck by the very sight of him.

Over the following days, Glenn returned to the Capitol to testify about what his flight had revealed, and he found much of his time consumed by signing autographs for lawmakers. He told them that his flight had proved humans were vital to space exploration because he had been forced to correct the capsule's orientation after technology had failed. He explained how each flight built on preceding missions. Glenn stressed the vital need for Congress to finance many more flights. Legislators listened and asked for more autographs. In the years to come, Congress continued to finance the hugely expensive space program. Measured in today's dollars, the cost of the rush to the moon would have amounted to each American contributing more than $200 a year to the space program.[22]

Perhaps the most widely reported part of Glenn's testimony was a declaration that had nothing to do with the importance of having a human being at the controls of a space capsule or with the once-doubted human capacity to work in a weightless environment. Republican senator Alexander Wiley of Wisconsin asked Glenn what thoughts he had in flight that related to his religious beliefs. Glenn replied, "I can't say that while in orbit you sit there and pray, or anything like that. It is a very busy time. . . . I feel that to me at least, my religion is not a 'fire engine' type of religion and not something that I call on in an emergency and then put God back in the woodwork in his place at the end of a particularly stressful period."[23]

Glenn's response attracted attention both inside and outside of religious communities, but the nation had little time to contemplate Glenn's philosophical ideas with the biggest day of the celebration yet to come. In New York City, an estimated four million people[24] bumped, wiggled, and shoved their way onto the sidewalks or hung out office windows as a record 3,474 tons of confetti and ticker tape streamed down from Manhattan skyscrapers.[25] Church bells tolled; bands marched. The noise was sometimes deafening, and the downpour of paper was sometimes blinding.

Faced with this welcome, John Glenn, the man who had seen four sunsets in a single day, was suddenly overwhelmed by events on planet Earth. He still had the good manners to shake hands and say a few kind words. Nonetheless, he was befuddled, bewildered, overwhelmed. As Newsweek reported, one young girl along the parade route shouted triumphantly, "I touched him! I touched him!"[26] Again, there were

The Glenns ride in New York's ticker-tape parade alongside Vice President Lyndon B. Johnson. *Courtesy of the Ohio Congressional Archives*

tears, tears of joy unleashed by simply seeing him. To many, he represented everything that was great about the United States; to him, crowd's response was beyond amazing.

The Glenns (and the other six astronauts and their families) first rode from LaGuardia Airport to city hall. Well-wishers lined the highway through Queens, hung from overpasses, and packed the side-walks along Broadway. Then, in a chaotic scene, New York City mayor Robert Wagner welcomed the astronauts to an outdoor ceremony at city hall. Glenn told the crowd, "Perhaps the efforts we're engaged in do really begin a new space era."[27] As he did in almost every appearance, Glenn credited the unseen thousands who had contributed to his success. As they celebrated, Americans did not see engineers, technicians, mathematicians, aircraft workers, or radio operators; they saw only him, smiling confidently and speaking with clarity and insight. The lord mayor of Perth, Australia, was among the guests on the podium. He had flown 11,624 miles to remind the world that his city had provided a nightlight for Glenn when he was on the world's dark side. After a few introductions and brief speeches, the cavalcade moved along.

Back in their cars, the astronauts waved as frantic police officers struggled to hold back the undulating and swelling crowd. In places, officers clasped hands to block the curious onlookers. Annie Glenn seemed most captivated by the people up in the windows, so high above and still members of the party. Annie liked parades better than other public events because they didn't require her to speak and possibly expose her stutter.[28] In the crowd, youths held a bedsheet banner that said, WE WANT GLENN 4 PRESIDENT.[29] *New York Times* photographers captured two unusual pairs of eyes in the sky—a window washer hanging above the crowd to watch the parade down Broadway and a teenager perched in a tree near City Hall Park.[30] For the people on the streets, getting a clear look at Glenn was like trying to spot a friend in a crowded football stadium. They sidestepped, maneuvered, and even jumped in hopes of getting a good view of the man of the hour.

Eventually, the crowd lunged past the barricades, and the parade stalled for a time. *Newsweek* estimated that "fully half of the wildly eager people . . . were frankly truant from schools in a three-state area."[31] Finally, the long, tumultuous ride ended with its arrival at the Waldorf Astoria Hotel, where there was a luncheon for more than two thousand people, a meal at which Glenn would have little time to eat; he would spend mealtime greeting celebrities who wanted to shake his hand. Mayor Wagner presented New York City medals of honor to all of the astronauts.[32]

Looking at the crowd of VIPs attending the luncheon, CBS News' Harry Reasoner reported:

> It's hard to think of another kind of dignitary, another kind of hero, that would attract the same kind of audience because they're not only the normal people like the mayor and Vice President Johnson, and Governor Rockefeller, who are going to be in this room. But there are representatives of all the major religious faiths. There are people from the UN. Circulating around here right now . . . some people from the entertainment world, some people from the business world, all here on short notice. They were selected as a cross-section of the big people in New York who should meet Colonel Glenn.

For a while, the TV camera showed the grinning Glenn shaking hands, trying to move a receiving line along swiftly. "Watching Colonel Glenn meeting people, most of whom must be just bland names, he retains an easy friendliness he might have at home on a Saturday night in Arlington, Virginia," Reasoner said. "I don't think anybody ever saw a man who became a hero so fast take it easier."[33] As much as this homecoming must have seemed to Glenn like moments from a Friday night episode of *The Twilight Zone*, Reasoner was right; Glenn did appear to be almost comfortable in his new exalted role.

On the following day, Glenn and his colleagues spent time at a United Nations reception. Even this usually staid organization followed the tone of the rest of Glenn's victory lap by greeting him with cheers. Soviet UN ambassador Valerian A. Zorin welcomed the astronauts enthusiastically, shaking their hands. Even Soviet premier Nikita Khrushchev, America's bogeyman of the moment, had sent Kennedy a congratulatory telegram with a surprising call for cooperation between the two nations. "If our countries pooled their efforts—scientific, technical, and material—to master the universe, this would be very beneficial for the advance of science and would be joyfully acclaimed by all peoples who would like to see scientific achievements benefit man and not be used for 'Cold War' purposes and the arms race," Khrushchev had written, raising hopes for joint space efforts in the future.[34] At the United Nations, Zorin reinforced that message, and that would lead to serious negotiations, although the first joint mission was more than a decade away.[35]

In political circles, Johnson received some criticism for trying to catch a glimmer of reflected glory by staying close to Glenn. In particular, his presence in Glenn's car during the ticker-tape parade prompted joking comments about the vice president, who smiled and waved from the passenger seat. The *New York Herald Tribune* published an article that criticized Johnson for being "that persistent rider on Glenn's shirttails."[36] Nevertheless, Johnson's mail from this period shows several letters thanking him for escorting Glenn without trying to capture the spotlight.[37] President Kennedy had, after all, named him to head the National Aeronautics and Space Council. When asked, Glenn had no complaints. "We had a very pleasant time with [Johnson and wife Lady Bird] every time we were together. . . . I appreciated his thoughtfulness and the president's in taking the time and the effort to be with us on those occasions. I thought it was fine myself."[38] More than twenty years later, Johnson's postmaster general, Larry O'Brien, was reminded of the controversy, and he said, "I don't think it was the only time he was accused of grandstanding. He had a style and a flair that lent itself to critics' fault finding."[39] Johnson, who felt stifled in the vice presidency, never turned away from a moment in the public eye.

Perhaps the strangest element of New York City's grand celebration was what many participants, viewers, and listeners overlooked. That morning, ninety-five people—eighty-seven passengers and eight crew members—had died in an American Airlines crash near Idlewild Airport (today John F. Kennedy International Airport). Bound for Los Angeles, the 707 took off successfully, but within minutes the automatic pilot system malfunctioned. The plane rolled over and crashed belly-up into Jamaica Bay. The airliner was full of fuel, so the impact caused a huge explosion that shook neighboring houses. Though it was clear almost immediately that there were no survivors, hundreds of police officers and firefighters spent the day gathering evidence for federal investigators.[40] On another day, an airliner crash would have been the talk of the town. This accident received less attention as the media, like the people in the crowds, focused on Glenn. In fact, many in the crowds were unaware of the crash. The next morning, the *New York Times* treated the two stories equally at the top of page 1. At that time, the crash was the worst single-airliner accident in US history.[41]

With his New York trip over, John Glenn could go home to celebrate. Another surprise awaited him there: when he arrived in New Concord, the village of fewer than two thousand residents produced a crowd of seventy-five thousand.[42] The college from which he never graduated, Muskingum, hosted sixteen hundred people in its gymnasium. Following a small-town kind of logic, organizers sent two invitations to every house on the local gas meter list. Ready to relax at home and just be Bud, the boy townspeople had watched grow up, Glenn learned that his name would now belong to a new high school, a portion of a state highway, and the very gymnasium in which he stood. Bud was forgotten as residents celebrated the triumph of their favorite son, Lieutenant Colonel John H. Glenn Jr.

Ultimately, Glenn was a bit embarrassed; he knew that the national media had overblown his youthful successes as a star athlete and stunning student, though he had been neither. He looked into the audience of familiar faces and said, "I know you teachers were pretty surprised to read about the straight-A average some newspaper writers gave me. . . . And you coaches who let me warm quite a few benches now discover you played all your key moves through me." With a grin, he admitted, "Coming home can cut you down to size."[43] Nevertheless, he ended the evening on a humorous note; looking around at what was now the John Glenn Gymnasium, he said, "Before you go home, be sure you clean up *my* gymnasium real good."[44]

John Glenn's travels had been a tour de force, from liftoff through the last celebratory events. *Time* began an article entitled "Colonel Wonderful" by saying, "Every so often a nation produces a genuine hero, raised above the multitude by acts of valor or virtue in times of war, crisis, or national frustration. He may come from any walk of life, so long as he fills the nation's need to elevate its vision and swell its pride. From Washington to Sergeant York, from Lindbergh to MacArthur, the U.S. has had its share of heroes. But few have encountered the universal approval and adulation that last week engulfed Astronaut John Glenn."[45]

And *Life*'s Picture of the Week made it clear that the celebration had touched lives across the nation, not just along the parade routes. The photo showed another somewhat bald head that otherwise bore no resemblance to Glenn's countenance. It was the face of a newborn boy named Lamar Orbit Hill, born in Ogden, Utah. The magazine also reported the births of "John Glenn Davis in Corpus Christi, Texas; John Glenn Tinsley in Cherry Point, North Carolina; John Glenn Eller in Gaylord, Kansas; John Glenn Donato in Burbank, California; John Glenn Apollo in Minneapolis; John Glenn Friend and John Glenn Nelson, both in Denver; John Glenn Anderson and John Glenn Garver, both in Salt Lake City"; and John Glenn Garnica in Dallas. (The Garnicas had planned to name their son Robert Kennedy Garnica when events interceded.)[46] And *Life* just scratched the surface. Newspapers across the country reported crib after crib filled with little John Glenns. And innocent tots were not the only ones taking Glenn's name. The astronaut received a magazine clipping that described three striptease artists calling themselves Alana Shepard, Gussie Grissom, and Jonnie Glenn. They billed themselves as "out of this world."[47]

"Next to John Kennedy," Tom Wolfe wrote in *The Right Stuff*, "John Glenn was probably the best-known and most-admired American in the world."[48] Wolfe also had a theory about why a glimpse of Glenn could generate tears in so many Americans, from the leaders of Congress to a scraggly teenager in New York City: It was "the aura, the radiation of *the right stuff*," of course. It was "the same vital force of manhood that had made millions vibrate and resonate thirty-five years before to Lindbergh—except that in this case, it was heightened by Cold War patriotism, the greatest surge of patriotism since the Second World War."

Friendship 7 also had its own fans. A newspaper photo from the weeks after Glenn's flight shows not a snowman but a snow space capsule built by four youths in Barnesville, Ohio.[49] The snow engineers had prominently displayed the name, *Friendship 7*, and underneath it was an American flag.

Glenn's own capsule was a heavily desired object. The mayor of Houston, Lewis Cutrer, urged Johnson to have it delivered to Houston, where it would be proudly displayed near the new Manned Spacecraft Center (today Johnson Space Center).[50] Alas for Houston, the coveted prize went to the Smithsonian Institution. The capsule's original home was in the National Air Museum. The museum's name was changed to the National Air and Space Museum in 1966, and today the capsule resides in the museum's current home, which opened in 1976.[51]

And if Glenn thought the parades would mark the end of the celebration, he was mistaken. In the first six months after his flight, he received more than 350,000 letters from all over the world. The first hint of the coming deluge appeared when the Glenns returned to their Arlington home after his speech before Congress. On that day, there were three gray US mail bags on the living room floor. The Glenn family wrongly assumed these bags represented all the mail addressed to the astronaut. Before long, a mail carrier appeared at the door with the unsettling news that the local post office was holding about a truckload of additional mailbags. The family opened the first three bags and started poring through the mail. Heads of state, royal and otherwise, wrote some of the letters; most came from ordinary people who felt a connection to Glenn that had been cemented during his spaceflight.[52] The letters allowed Glenn for the first time to "see how the country had bonded, and how many others around the world had found common cause with the United States and its mission into space."[53] Soon the Glenns' carport was full of mail bags. One letter came from Anatoly Dobrynin, the Soviet ambassador to the United States, who asked Glenn to autograph US stamps issued in his honor. The ambassador explained that the stamps would be used in a children's show in the Soviet Union.[54]

Eventually, Steve Grillo at NASA took over handling the mail. He set aside letters that seemed special so that Glenn could see them, and Grillo sent replies carrying his own signature to the others. As of August 31, 1962, Grillo had sent out more than forty-one thousand replies to simple congratulatory letters as well as requests for auto-graphs and photos. NASA produced most of them automatically by using electric typewriters; however, each one had to be addressed individually. Grillo's team returned to senders any money or stamps enclosed in Glenn's "fan mail."[55]

Included in the total mail received were thirty-seven hundred messages from people in other countries who wrote in English and twelve hundred from foreign letter writers who scrawled or typed in other languages.[56] All in all, international response to the flight had been extraordinarily positive. Indeed, it was difficult to name a nation that had not responded enthusiastically. China probably came clos-est. Chinese leaders voiced no negative response, and yet the nation's media waited sixteen hours to report the flight, an unthinkable gap in the minds of some Americans. The Chinese also reportedly provided only "meager attention."[57]

Many letters came from children, with clever little poems, such as: "Roses are red / vetols [sic] are blue / when you went around the world / we did to [sic]."[58] What this child was telling Glenn was something new: by using the technologies of television and radio and by broad-casting Glenn's heartbeat and his voice, NASA had created a virtual spaceflight experience and a virtual community consisting of millions of Americans and others around the world. The letters struck Glenn as so fascinating that he eventually had the best of them published in a book, *P.S. I Listened Your Heart Beat: Letters to John Glenn*. The title sprang from a child's letter.

The parents of a Michigan boy sent Glenn their six-year-old son's prayer for him and admitted that he had a wee bit of trouble with names:

> Oh, God we thank you for John Glenn who went around our whole United States World three times. We are so very thankful his capsule didn't get on fire but that you brought him back safe to Cape Canaveral. We also thank you for bringing back safe Roger Shephard and, the other guy. Oh God we are thankful for the men the Russiantons sent up came back safe too! Amen.[59]

From an eleven-year-old Californian: "What happens if you are in a space helmet and your nose itches?" Glenn's response: "Pure torture, believe me! But you'd be surprised at how adept one becomes at nose 'wriggling' without having anything solid to rub against. Try it sometime."[60] Glenn's ability to make grown men cry apparently had no national boundaries. He received a letter from a captain in the Japanese navy who "felt the tears gather in my eyes" as he read the details of Glenn's flight.[61] Perhaps the most poetic letter came from an eighty-eight-year-old Briton who wrote:

> Like the rest of the world I was thrilled with your triple journey round the earth and offer you my sincere congratulations on so epic a deed. I am glad to read that you suffered no after-effects from the speed and altitude. You can in all truth say of yourself, like Hamlet, "I could be bounded in a nutshell, and count myself a King of infinite space."[62]

And Glenn was not the only person receiving mail related to his flight. Glenn fans also bombarded President Kennedy and Vice President Johnson with mail, and they added to Glenn's pile by forwarding their letters to him. Chet Hagan, producer of NBC News, wrote to thank Annie Glenn and to apologize if she experienced any problems caused by his network's employees. He gave no details about the incident that sparked his apology.[63]

NASA also fielded an avalanche of requests for appearances by Glenn. Invitations covered a wide variety of requests:

- to visit colleges and military bases
- to make speeches at graduation exercises and club luncheons
- to appear at an air show, an American Legion Loyalty Day Rally, a July Fourth celebration, and a naval recruiting rally
- to judge a science fair
- to help a high school marching band to raise money for a European tour
- to serve as grand marshal of a parade

- to receive an honorary degree
- to be interviewed by a school newspaper

These invitations came from various parts of the United States, and from as far away as the Netherlands and Switzerland. NASA made lists of the requests. One list included twenty-four invitations between April and July. Running down the right-hand side of the page was a column labeled ACTION, and beside every invitation on the list was the same word: DECLINED.[64] NASA deputy administrator Hugh Dryden estimated that requests for Glenn arrived at a rate of ten to fifty each day after his flight. Glenn's superiors had asked him whether he wanted to continue as an astronaut or to accept these invitations and become a full-time honored guest, Dryden said. When Glenn confirmed that he preferred a career in space, NASA decided to protect him as much as possible from these requests.[65]

At around the same time, General Foods wanted to put Glenn's face on a Wheaties box. The opening offer was $1 million, and the astronauts' agent said he was sure the cereal company would pay more, but Glenn said no. Despite his new status as a celebrity, the forty-year-old father could not picture his face on a cereal box. "I just didn't think it was right," he later wrote.[66]

There was one civic group that John Glenn happily supported: the Boy Scouts of America. Although he was unable to attend the first annual conference of Eagle Scouts in Ann Arbor, Michigan, he took the time to send a taped message to the scouts. Though he'd never been a scout, Glenn shared the organization's ideals. His son, Dave, was an Eagle Scout and his daughter, Lyn, a Girl Scout. He spent hours out of his day recording several messages in support of the Boy Scouts' "Go" campaign to expand scouting so that more boys could participate.[67] Glenn also remained grateful to the New York Council of Boy Scouts for giving up the grand ballroom at the Waldorf Astoria on New York City's "John Glenn Day." Even amid the crush of events on that frenzied day, Glenn had taken a few minutes to thank those scouts, who had moved their luncheon to the hotel's Starlight Roof.[68] In April, the National Executive Board of the Boy Scouts of America unanimously voted to give Glenn the Silver Buffalo Award in recognition of his impact on the youth of America.[69] The fall 1962 edition of *"Go" Roundup News*, a Boy Scouts publication, included three photos and one sketch of Glenn on the front page.[70]

Around the country, the Glenn-a-thon continued running in both official and unofficial circles. Adlai Stevenson, speaking to the Twelfth Annual Conference of National Organizations, called by the American Association for the United Nations in Washington, told the attendees: "Let us, therefore, be grateful for that image of *Friendship 7*, carrying round the earth one of the most buoyant and manly personalities, and one of the clearest, most light-of-day minds, ever 'orbited' into the national consciousness. For it has begun to replace some of the images of unreasoning fear to which we have been treated recently."[71]

On April 9, CBS News broadcast a special televised report on *Friendship 7*. This program, shown forty-eight days after the flight, aired previously unseen photographs and film footage of Glenn on the day of his flight. It included images of him inside the space capsule throughout the flight as well as his voice communications and his heartbeat. Glenn told Walter Cronkite, "I think humans always have had fear of an unknown situation. . . . The unknown areas have been shrunk to an acceptable level. . . . We've been looking forward to this." He also explained that his worries about the heat shield had been dampened somewhat by his powers of observation, which registered none of the banging sounds that he would expect if the shield had been loose.[72]

As the months went by, Glenn began to discover several downsides to being famous. The lack of privacy was obvious immediately; other problems soon arose. For example, he encountered hoax believers. This group of people fell into two groups: the first merely believed that Yuri Gagarin was an actor who had not actually gone into orbit; the second faction believed that Glenn had never flown around the world, and at least one believed that Johnson's smiling face during Glenn's address to Congress was a reflection of his knowledge that Glenn was manufacturing creative science fiction.[73]

Five months after his flight, Glenn got himself into long-lasting trouble with testimony before a congressional subcommittee. The

House's Special Subcommittee on the Selection of Astronauts was investigating the possibility of adding women to the astronaut corps. Supporters of women seeking spots on the astronaut roster cited several advantages female astronauts would have: they were smaller, a definite plus in a compact spacecraft; they outscored men on some mechanical exercises; and they outperformed men on some medical and psychological tests. At the time of his statement, a group of woman pilots, all with more than eight thousand hours in flight, had volunteered. Some of them already had more flying experience than some male astronauts. One woman, Geraldyn "Jerrie" Cobb, had completed three programs of thorough training, and twelve more had completed the first phase.[74]

While expressing a willingness to accept woman astronauts, Glenn repeatedly returned to a premise that excluded them: all astronauts needed the flying and engineering experience that could be gained only by being military test pilots. At that time, the US military had no female test pilots, so that restriction made it impossible for women to be astronauts. He feared that special training for women would add time and expense to the space effort. In the hearing, Glenn made a couple of statements that might have been acceptable to many in 1962, but angered others.

Looking back at those statements decades later makes them seem cringeworthy and terribly backward because the roles of women in American life have changed so much since then. For instance, when it was pointed out that some female pilots had passed the same physical exams that the astronauts had faced, Glenn said, "A real crude analogy might be: we have the Washington Redskins football team. My mother could probably pass the physical exam that they give preseason for the Redskins, but I doubt if she could play too many games for them."[75] To defend his position, Glenn tried to make a sociological argument: "I think this gets back to the way our social order is organized, really. It is just a fact. The men go off and fight the wars and fly the airplanes and come back and help design and build and test them. The fact that women are not in this field is a fact of our social order."[76]

As a man who would develop political ambitions, Glenn had

committed the crime of thinking inside the box and showing that he was not among the vanguard willing to consider dramatic change. Even at the time of his death in 2016, some feminists blamed Glenn for delaying women's participation in space exploration. Nonetheless, there is inherent unfairness in using today's social standards to judge statements made more than half a century ago.

There was also a brief flurry of bad publicity about one- and two-dollar bills stowed away in the *Friendship 7* capsule. This prompted a congressional investigation amid concerns that Glenn may have placed the bills there so that he could profit from selling them. When completed, investigations cleared Glenn. A NASA document revealed that there were two bundles of bills with a total value of fifty-two dollars to fifty-six dollars. The bills belonged to a launchpad employee of McDonnell Aircraft. Inspectors from both McDonnell and NASA approved their insertion in the space capsule.[77] Fifty-four years later, one of the one-dollar bills sold for $18,499.73.[78]

Despite occasional controversies, Glenn remained America's hero and a media sweetheart. In photos and in text, he scored points with readers. His all-American good looks, his humility, and his virtuous outlook combined to make him a heralded representation of clean living and American exceptionalism. Many asked what other nation could have produced this man. As Walter A. McDougall wrote in his 1985 work, *The Heavens and the Earth: A Political History of the Space Age*, not even the moon landing would generate the kind of "national catharsis" created by Glenn's achievement and its effect on Americans' view of the Cold War. It was, McDougall stated, a "social release into which John Glenn . . . incredulously stepped."[79]

His life was forever changed, but many who knew him over the years said that the man and his demeanor remained true to Bud, the small-town boy with a yearning to fly. As an astronaut, and later as a private citizen and a senator, signing his name to scraps of paper became a daily ritual. More accessible than a movie star or a US president, he continued to greet strangers as friends.

8

WHAT'S NEXT?

For John Glenn, the aftermath of *Friendship 7* became, in many ways, more complex than its prelude. The details of his mission had been mapped out in exquisite detail before he lifted off, but after it was over, this longtime military man and astronaut had no star chart to guide him through the treacherous territory awaiting anyone celebrated as a hero. After more than two decades of going where the Marine Corps or NASA had told him to go, he was forced to improvise and sample the many possibilities open to him in the wake of his accomplishments in space and his stunning victory tour. At times, he stepped forward boldly; at others, he faltered and hesitated on the brink of his unmapped future. Glenn wanted to stay in the space program and go to the moon, but he knew there were no guarantees.

After his flight, the Kennedy administration gained momentum in its push to reach the moon by the end of the decade. NASA's new "Long Range Plan" reached the White House in May and reflected "a major realignment to match the moon goal" as well as JFK's other stated priorities—communications satellites, meteorological satellites, and exploration of the use of nuclear-powered rockets. All of these goals were intended to guarantee that US science and technology would not become outdated, to "provide insurance against the hazards of military surprise in space," and to take a world-leading position through the space age.[1] Another draft document by executive secretary to the National Aeronautics and Space Council Edward C. Welch stated, "Because of the revolutionary nature of space technology, the achievement of national objectives can wait neither for perfection, nor for slow-paced development."[2] The document went on to say, "There are inherent risks in space exploration, particularly in manned flight, and they must be recognized. A basic element of the national space program is accepting those risks, taking reasonable safeguards, and absorbing the unavoidable losses which will occur."[3] In other words, the astronauts were expendable—and they knew it.

As the United States planned a thorough reconnaissance satellite program, Kennedy issued National Security Action Memorandum 156, which urged government officials involved in treaty negotiations to proceed "with a view to formulating a position which avoids the dangers of restricting ourselves, compromising highly classified programs, or providing assistance of significant military value to the Soviet Union."[4] Clearly, reconnaissance was another priority, though the United States wanted to maintain as much secrecy as possible about its espionage capabilities.

While Cold War planning continued, Glenn was happy to have an opportunity to break through the wall separating the US and Soviet space programs. He played host to Soviet cosmonaut Gherman Titov in May when both attended the Committee on Space Research of the International Council of Scientific Unions' third International Space Symposium in Washington. Beyond more formal and official events scheduled around the symposium, John and Annie Glenn had invited Titov and his wife to their home for dinner at 6:30 one evening. However, the two couples received invitations to a reception at the Soviet embassy on that night, and the Glenns assumed that their dinner plans had been canceled. Much to their surprise, as they were nearing the end of a long receiving line, a State Department employee told them that the Titovs had accepted the invitation to have dinner at their house in thirty minutes.

Louise Shepard, Alan Shepard's wife, joined John and Annie Glenn as they rushed home to make last-minute preparations for an acceptable dinner. NASA asked the Arlington police department to provide a motorcycle detail to reinforce security, and the hurried hosts immediately dispatched the police officers to visit a convenience store to buy frozen peas. The Glenns borrowed frozen steaks from neighbors and started a fire in two charcoal burners in their carport. Glenn threw the steaks on the grill, while the two women kicked off their dress shoes and worked feverishly in the kitchen. Suddenly, the fire grew too hot, forcing Glenn to dash inside for water. When he returned, the flames were three feet high. There was so much smoke pouring out of the carport that it looked like the house was ablaze. Then a horrified Glenn saw that the "little forge that holds the grill on top of one of the burners broke and dumped a load of steaks in the fire." He tossed more water on the grill, creating a cloud of steam to join the smoke. At exactly that moment, the Titovs arrived, and Glenn told them in a matter-of-fact way that if they expected food for dinner, they would have to help. Titov's wife, Tamara, headed for the kitchen, where she shucked her shoes and started to work on a salad; her husband assisted Glenn with the steaks. The astronaut said that he thought "we had probably set the cause of diplomatic relations back about 50 years." Hearing an account of the nearly disastrous dinner party later, President Kennedy "really got a kick out of it," Glenn said.[5]

Glenn and Titov also appeared on a May 6 television program. NBC broadcast a panel discussion as part of the series *The Nation's Future*. In addition to the two space travelers, the participants were Hugh L. Dryden, NASA's deputy administrator, and Anatoly A. Blagonravov, a Soviet space scientist and vice president of the Committee on Space Research. Without being totally revealing, the discussion, led by NBC's Edwin Newman, filled some information gaps. For example, Titov admitted that the rumors were true: he had suffered debilitating motion sickness throughout the weightless portion of his flight. He said that he never had a similar experience while riding in a car or a

Cosmonaut Gherman Titov, left, and Glenn speak to reporters on May 3, 1962, in Washington. *John F. Kennedy Library and Museum, Boston*

plane, so "perhaps my organism has this peculiarity of not resisting spaceflight sufficiently well."[6] Dryden identified astronauts' reactions to weightlessness and radiation as the two potential trouble areas that concerned NASA.[7] Glenn added that NASA must focus on both the immediate effects of weightlessness, such as Titov's illness, and the long-term possibility of bodily deterioration.[8]

As part of the discussion, Glenn mentioned that he had experienced equipment failures during his flight, and he asked about malfunctions encountered by Soviet cosmonauts. Titov denied that cosmonauts had experienced a single problem in space: "I can tell you quite frankly that there has been no failure in manned flights in the Soviet Union at all. There have been no failures whatever." Titov may have been following a party line, but he may have believed that the Soviet space program had not encountered a single malfunction, as unlikely as it seemed. Soviet flights of this era were controlled almost completely by technicians on the ground and involved little cosmonaut activity, so he may not have been aware of problems. He went beyond answering Glenn's question to attack American reports of failed Soviet flights: "There is a commentator in your country by the name of Drew Pearson whose articles create misapprehensions and false impressions about alleged failures by Soviet Cosmos ships. I have read his essay in which he alleged that a number of cosmonauts preceding Gagarin died in the course of descents which failed." Titov contended that such deaths would be impossible to hide, adding, "After all, even a cosmonaut has a mother and a wife and let's say the mother lost her son. . . . Everybody will find out by the grapevine."[9] American stories of cosmonauts dying in space were rampant among the public during this era, and Pearson merely added fuel to the fire.[10]

A French journalist asked Glenn and Titov, both handsome men, "how it feels to be suddenly and generally a hero, such as [American actor] Gary Cooper or [Ukrainian actor and director] Sergei Bondarchuk." Glenn and Titov had already discussed this issue; therefore, Titov answered for both men, saying they had plenty of technical training, "but there is no training at all in running press conferences, in making public appearances. So you see this is tough and heavy work for us, for which we are not trained."[11] Dryden reported that conference participants had discussed greater cooperation between the two space programs and hoped to negotiate details in Geneva.[12] Titov said that in science, cooperation was vital.[13]

After his visit to America, Titov spoke about Glenn's "tenacious professional gaze" and said that the two men were "connected by everything they had experienced and lived through in space."[14] Nevertheless, Dryden's talks with Soviet officials about cooperation in space led to no major partnerships.

In 1962, NASA was on the move on the ground as well as in space. The agency's relocation of its base management and planning operations from Langley Research Center in Virginia to the Manned Space Center (today Johnson Space Center) in Houston forced the astronauts and others working in the manned space program to move to Houston. As they faced the unavoidable upheaval, the astronauts confronted an unexpected controversy. The Houston Builders Association offered to provide free homes for the seven Mercury astronauts. After a public uproar, public affairs officer "Shorty" Powers announced, "NASA has determined that there is no legal bar to acceptance of the offer. As a matter of policy, however, NASA has advised them that acceptance of the houses is not considered to be in the interest of all concerned." NASA still left the final decision in the hands of the seven astronauts. The astronauts' agent and attorney, C. Leo DeOrsey, advised them in April that it was best not to accept the houses, and they agreed.[15]

Later in 1962, the Mercury astronauts and the New Nine, the second class of space explorers chosen in September 1962, began moving their families to Houston and settling in homes close together, establishing spaceman colonies. John and Annie Glenn and Rene and Scott Carpenter decided to build new homes in the Timber Cove development with neighboring lakeside lots. The Schirras and the Grissoms landed nearby. In July, Houston held a parade and Texas barbecue fiesta welcoming the Mercury astronauts to town. As soon as they had moved in, tour buses appeared outside their homes loudly announcing who lived in each house.[16] Settled in their new home, the Glenns threw a party to welcome the New Nine and their spouses.[17]

In the early summer of that year, Attorney General Robert Kennedy invited the Glenns to join him for a weekend at the Kennedy compound in Hyannis Port. While sitting on the fantail of the

Kennedy yacht, the *Honey Fitz*, President Kennedy questioned the propriety of the astronauts' exclusive agreement with *Life* magazine. Glenn explained that the agreement covered only their private lives and involved no sharing of information about the space program. Glenn said he'd be happy to give back the money he had received if anyone could show that he had withheld any official information from the rest of the media.[18] Later, he was widely credited with convincing Kennedy to allow the initial deal to remain in place and opening the way for the New Nine astronauts to join a similar arrangement.

In September, astronaut Wally Schirra caused a brief furor with claims that Glenn had not been able to maintain his training schedule as a result of the public relations assignments piled upon him. Both Glenn and NASA administrators denied these charges, while conceding that he had been in great demand. "All of Glenn's appearances have been carefully screened and his activities have been the same as the other six astronauts," Powers said.[19]

In truth, Glenn was frustrated by his long schedule of public appearances. He visited Webb's office one day and told him that he no longer was willing to crisscross the nation, appearing at events just to make an individual lawmaker happy. Webb claimed that Glenn was only sent on trips that were important to strengthening congressional support for the space program. Glenn stated unequivocally that he was unwilling to continue these appearances. Despite Webb's obviously angry response, the astronaut refused to capitulate. As Tom Wolfe wrote, Glenn "made it obvious who held the cards around here, and that was that."[20]

Meanwhile, US spaceflights continued to expand NASA's knowledge and experience. Deke Slayton had been scheduled for the flight after Glenn's, but he was grounded by physician concerns about a heart murmur. He soon became NASA's coordinator of astronaut activities. In his place, Carpenter made three orbits in May. He earned the ire of Chris Kraft, who thought the astronaut had overused his ship's fuel as he enjoyed sightseeing through the capsule window. Splashdown occurred 250 miles off target after Carpenter had to manually fire the retro-rockets and ran out of fuel during the descent. Kraft swore that this astronaut "would never again fly in space"—and he didn't.[21] Next, Schirra made six orbits in a near-perfect October flight. NASA also launched the first commercial satellite—Telstar—in July 1962. Designed by Bell Telephone Laboratories, the communication satellite and its successors that bore the same name would become identified with international television broadcasts.[22]

For John F. Kennedy, manned space travel had meaning primarily as a way to beat the Soviet Union to the moon and claim victory. In November 1962 a new debate arose in the White House about that goal. At issue was a proposal to initiate a crash program to get men to the moon by 1967 before the end of an anticipated second term for Kennedy. Both Secretary of Defense Robert McNamara and Webb spoke out strongly against acceleration. Webb argued that there were "real unknowns to be resolved to reach the moon." Kennedy contended that "everything that we do ought to be tied in to getting to the moon," admitting, "I am not that interested in space."[23] The specificity of the moon goal appealed to the president, heightened public interest, and boosted congressional support and funding in a way that no subsequent endeavor has. From a White House perspective, sending men to the moon offered the romance of exploration, but the mission was powered by Cold War politics as much as it was by rockets.

The Mercury space program ended in May 1963 with Gordon Cooper's twenty-two-orbit mission, which lasted more than thirty-four hours. He was the first astronaut so relaxed that he fell asleep during the countdown. Astounded flight controllers had to wake him several times during the last hour before launch. He also slept during the flight, which was appropriate, given its length.[24] Glenn's role in this flight was to communicate with Cooper from a tracking ship between Honshu and Okinawa. Toward the end of Cooper's flight, a series of problems developed. Over the course of a few minutes, his instruments were incapacitated by electrical surges, and the capsule's automatic system stopped working. When mission control checked Cooper's condition, he replied in his easily identifiable Oklahoma twang, "Well, it looks like we've got a few little washouts here. I've lost all electrical power. Carbon dioxide levels are above maximum limits, and cabin and suit temperatures are climbing. Looks like we'll have to fly this thing ourselves. Other than that, things are fine."

Slayton laughed aloud at Cooper's seemingly unworried account and ordered mission control to make his splashdown on time, "not a second later." As a result of the automatic control system failure, Cooper had to perform retrofire manually, and Glenn gave him instructions for the procedures worked out at mission control.[25] Glenn "actually counted him down to his time when he pushed the button up there himself."[26] He reassured Cooper, telling him, "It's been a real fine flight, Gordon. . . . Beautiful all the way."[27] Cooper splashed down safely almost exactly where he should have.

With the Mercury program ended, Glenn began preparing for the third manned flight program—Project Apollo—which would take men to the moon. He was assigned to "pioneer" the overall effort for a lunar landing.[28] Specifically, he was part of a group overseeing design of the lunar module, which would land on the surface of the moon. Grumman, headquartered in Bethpage, New York, was constructing the spacecraft.[29] Other astronauts focused on the second space program, Project Gemini, which would feature two-man vessels, with return flights by some Mercury astronauts and first missions for the New Nine. Gemini would attempt intermediary steps that would be necessary to reach the moon, such as making space walks and maneuvering to master space rendezvous and docking.

NASA assigned New Nine astronaut Pete Conrad to lunar module development as well. Conrad traveled with Glenn, who was routinely mobbed by autograph seekers. They tramped through airports, with Conrad carrying both bags so that Glenn's hands were free to wave, shake hands, and sign his name. Conrad "looked like some little guy who carried bags for John Glenn. What was more, that was what he felt like," according to Wolfe.[30]

A list of astronaut public appearances for the first seven months of 1963 showed Glenn making four appearances, all at national conventions. Grissom participated in an Air Force Academy event. Carpenter made four appearances—things like the dedication of a park and swimming pool in Boulder, Colorado, or a presentation at the West Virginia Centennial Science Honor Camp. Schirra made five, including presentation of Glenn's capsule to the Smithsonian, and Cooper attended two events. Even some of the new guys—Conrad, Jim Lovell, John Young, Tom Stafford, Frank Borman, Elliot See, and Ed White—made appearances at less prestigious events, such as year-book dedications and assembly programs.[31]

Glenn remained the media star. In January 1963 his mother authored a cover piece for *Guideposts*, a nonprofit religious magazine. In it, she wrote about his youth in New Concord.[32] Three months later, excerpts from a speech at the Associated Press's annual meeting were released in the form of an AP spiritual piece by Glenn, who asserted that the nation's success "was woven of belief—belief in God."[33] He wrote "Book Larnin' Ain't Everything," an article for *Utah Education Review*, in May 1963. Encouraging educators to kindle students' imaginations, the astronaut argued, "As we educate, we place many facts into the minds of young people. . . . We answer the question of why things are as they are. However, we must relate experience to the future and answer the question of how we can best apply such information."[34] US representative Ken Hechler of West Virginia argued on September 16 that a speech Glenn had delivered to the National Conference on Citizenship "articulates what goes to make up the American dream."[35] A month later, the *Long Island Press* reported that the Kennedys were looking to attach the president's reelection campaign "to the soaring moon-bound star of Astronaut John Glenn Jr."[36]

All the astronauts profited from contact with the press. A new deal, encompassing all sixteen, promised each one $16,250 a year for four years from *Life* and Field Enterprises, both of which bought the rights to their personal stories. *U.S. News & World Report* examined the pros and cons of the deal[37] and then asked Glenn to speak on behalf of the astronauts. He politely griped about the loss of privacy suffered by all astronauts. "I never thought of myself as a candidate for an unlisted phone number," he said. "But what do you do after the umpteenth call in the middle of the night—especially from the drunks who want to prove to friends that they can get through to you? Inevitably, you must shut it off." He argued that this contract offered a measure of privacy by limiting media intrusions into their personal lives.[38]

During this period, Webb repeatedly met with JFK about the expense of the space race. He told the president that NASA needed more than $5 billion in both 1964 and 1965. Kennedy replied that the administration must tie the expense to the Cold War to make it justifiable. Webb asked the president whether funding the program would be easier with a military man at the head of NASA. Kennedy did not embrace that idea, although he concluded that a "national security" tag was essential.[39]

In October 1963 the two classes of astronauts endured jungle training in Panama to prepare for unexpected landings in tropical locales. Glenn, who had just returned from a tour of Japan, joined his colleagues on the four-day survival course. The curriculum for this training program included lessons on how to hack through jungle vegetation and how to kill and consume iguanas, monkeys, and insects. After some basic instruction, the astronauts entered the jungle in three groups with two instructors attached to each group as observers, not helpers.[40] Neil Armstrong was Glenn's training partner. Each man hung a hammock, and Armstrong jokingly hung an extra one just in case a Chocó wanted to join them. Because they were in Chocó territory, they had been alerted to possible encounters with the native peoples. They did see one man wearing paint. "I never did know whether he put it on just for our benefit or whether that's the way he normally ran around," Glenn said later.[41] He was amused by Armstrong's offbeat attitude about exercise—something Glenn took seriously. Armstrong contended that each person was born with a certain number of lifetime heartbeats and swore that he did not want to consume his by doing "something silly like running down a road," in Glenn's words.[42]

Always interested in exchanging information with his Russian counterparts, Glenn wrote a memo to the White House on November 4, 1963, proposing that he work to establish means for information exchanges between the two nations.[43] In a separate proposal, he advocated greater sharing of information with other countries through establishment of a Manned Space Flight Scientific Briefing Team. Both recommendations were presented for President Kennedy's consideration.[44] Four days later, Webb wrote to Dryden and associate administrator Dr. Robert C. Seamans Jr. that Glenn had been told he might be pulled from NASA duties to participate in "extra-astronaut" activities, such as those he had proposed.[45]

Meanwhile, Glenn had become increasingly frustrated by not seeing his name on the list of astronauts assigned to upcoming flights. "I wanted to get back in rotation and go up again," he said. "Bob Gilruth who was running the program at that time, said that he wanted me to go into some areas of management or training and . . . I said I didn't want to do that. I wanted to get back in line again for another flight. But he said headquarters wanted it that way [with Glenn off the flight schedule], at least for a while."

Years later, Glenn learned the real reason for his grounding: "Apparently, President Kennedy had said that he would just as soon I wasn't used again for a while. . . . I guess after my flight there had been such an outpouring of national attention. . . . I guess adulation is the word that comes to mind."[46] Walter D. Sohier told Glenn in 1964 that President Kennedy had "thought it would be a great mistake if you were used on any future flights, because you were such an important asset to this country." Sohier believed Kennedy was referring to Glenn's "inspirational aspect."[47] Looking back, Glenn said, "Obviously, I would have liked to go to the moon . . . but I had no complaints regarding my role in the space program."[48] In this case, his status as a hero had limited his options at NASA while opening up new opportunities elsewhere.

In mid-1963, Robert Kennedy had approached Glenn and asked him whether he would be interested in running for political office, such as the US Senate seat up for grabs in Ohio. Glenn's initial answer was that he might pursue public office someday, but the timing didn't seem right. He changed his mind after President Kennedy was gunned down by Lee Harvey Oswald on November 22, 1963. In the assassination's wake, Glenn heard a more urgent call to public service. (He and Annie were the only astronaut couple invited to attend JFK's funeral.)⁴⁹ "So I went back to Bobby again," he said, "and told him that I had changed my mind."

The attorney general feared that Glenn's decision had come too late to put together a successful campaign against Democratic incumbent Steve Young.⁵⁰ Bob Voas, a member of the Mercury astronaut training program, became Glenn's campaign manager. When the astronaut announced his candidacy, "we had no organization at all," he said, although he did have some influential backers within Ohio's Democratic Party.

While Glenn could leave NASA whenever he chose to depart, retiring from the military was more complicated; he had to expend forty-five to sixty days of leave time and navigate a paperwork jungle. Under military regulations, he was not allowed to engage in active campaigning until he completed his Marine Corps service.⁵¹ With Glenn entering a primary battle against Young, the State Democratic Convention remained neutral, and President Lyndon Johnson chose to avoid an intraparty fight as well.⁵² Young didn't hesitate to attack Glenn as a political novice. "There were statements along the line of, I was last in the line of chimps who had been in space," Glenn said.⁵³ The former astronaut also faced criticism from some Americans for leaving the space program. "I think at that time we'd been put on a pedestal by a lot of people," Glenn acknowledged.⁵⁴

When he went to Ohio to begin preparations, the space hero fell in a bathroom and suffered a concussion and other injuries. He developed an inner ear problem that discombobulated him, ruining his sense of equilibrium. His ailment, "traumatic vertigo," might cause lifelong trouble, but nine months to a year of treatment could likely avert any lasting consequences. He was initially admitted to Grant Hospital in Columbus and then sent via medevac to Milford Hall Hospital at Lackland Air Force Base in San Antonio. Glenn had not officially left the Marine Corps, so he received military health care, although he had been reassigned and was no longer tied to NASA.⁵⁵

During his treatment, Rene Carpenter and Annie decided to campaign in Ohio and "try to keep this alive," in Glenn's words.⁵⁶ When Glenn faced the reality that he could not travel and campaign during his recuperation, he decided that the only way he could run would be solely on his record as an astronaut, and consequently, he dropped out of the race. The campaign owed $16,000 when he withdrew. In his memoir, Glenn quoted his then-teenage daughter as commenting on his fall and its aftermath, saying, "That may have been the first time that I realized my father was human."[57] Months later, *New York Times* writer James Reston teased Glenn: "Here you fly about the universe and then you skid in a bathtub and knock yourself out. This must have been quite a year, wasn't it?"[58]

While he recovered, Glenn considered his options, including a return to NASA.[59] Reflecting on the possibility of a career at the space agency, Glenn said he did not want to spend the rest of his life as a training officer or to become nothing more than a "used astronaut," wrote fellow astronaut Walter Cunningham.[60] He remained popular and won the Freedom Foundation's top award, the George Washington Medal. Several Democrats recommended Glenn to Johnson as a potential running mate in his 1964 reelection campaign or as an official in the Johnson administration.

Glenn also received offers from various businesses. He hoped to hear from a company that showed an interest in "my coming with the company in a business sense to develop and be on the board, and do things like that and make decisions," he said. "Royal Crown [Cola] was the only company that would do it that way."[61] In August 1964, he accepted a $50,000-per-year contract at Royal Crown. Under his contract, he would represent the company as a goodwill ambassador and board member, while working as a NASA consultant and an active supporter of both the Boy Scouts of America and the Freedom Foundation.[62] Family life was changing too. Both Dave and Lyn were in college by the fall, and John Glenn Sr. lost a six-year struggle with cancer and died.

In October 1966, Royal Crown promoted Glenn to board chairman of Royal Crown Cola International, which was undergoing a major expansion that included new operations in Ecuador and Thailand. The international headquarters moved from Fort Lauderdale to New York, and the company planned new world field offices.[63] In addition, Glenn would lead an "immediate five-country European inspection tour" in search of new growth opportunities. This trip would be followed by efforts to establish franchises in the Middle East, Africa, and the Far East.[64] "Annie and I then had an apartment in New York and spent most of two years traveling pretty much all over the world, in places where there were contacts about people who might want to franchise Royal Crown," Glenn said.[65] "They were paying me $100,000 a year, which at that time was real money. For the first time in our lives, Annie and I didn't have to worry about putting our kids through college or helping our parents financially as they got older."[66]

On NASA's behalf, Glenn made a five-week European tour in 1965 as well as a nine-day trip to Burma and a month-long tour of Europe in 1966.[67] US ambassador to Belgium Ridgway B. Knight issued a glowing report on the Glenns' private lunch with King Baudouin in May 1966. Calling it his "dream," the king had asked Glenn whether he could attend a spaceflight launch. The former astronaut thrilled him by promising to make the arrangements. "At no time during my 11 months in Brussels has there been so much favorable publicity for the U.S. over radio, television and in press as during the three days of the Glenn visit," Knight reported to Washington.[68] On Glenn's next stop—France—the public reception was so frenzied that he reported "if there had been any more enthusiasm, in some instances, someone would have gotten hurt." A crowd of thousands greeted Annie, John, and Lyn Glenn, and the astronaut-turned-cola-franchiser became the second honorary citizen of Paris. The first, the Associated Press reported, had been World War II general George S. Patton, who was honored for his decisive role in routing German forces from France and contributing to the Nazi defeat.[69]

In January 1967, NASA suffered its first loss of astronauts in a spacecraft.[70] The Apollo 1 crew—Gus Grissom, Ed White, and Roger Chaffee—died in a launchpad accident during a test at the Cape Kennedy Air Force Station. An electrical short set off a fire that spread rapidly in a pressurized, pure-oxygen atmosphere. The astronauts were trapped by a hatch that had not been designed for quick escapes. An investigation identified many changes that would be necessary before the moon mission could be set into motion again.

Later in the year, Lyndon Johnson tried to lure Glenn into his administration as head of the President's Council on Physical Fitness. After lobbying Glenn, Webb reported that he might have to be forced into the job. Johnson felt Glenn should be honored to take the post. LBJ's only reservation about Glenn, which he expressed in a telephone conversation with Webb, was that he "flirts with Bobby [Kennedy] some."[71] At this time, Johnson and Kennedy were archenemies.

Months later, Glenn took a shot at being a TV star. Producer David Wolper recruited him for a proposed new documentary series that would trace the journeys of previous adventurers. As an explorer himself, Glenn seemed like an ideal host. His first assignment for *Great Explorations* was to track the route taken by journalist Henry Stanley during his nineteenth-century search for the missing Scottish missionary David Livingstone, who had been almost totally out of contact with the outside world for three years after he left for central Africa in 1866 on a quest to find the Nile's source. Stanley, a correspondent for the *New York Herald*, set out in 1869. He arrived in Zanzibar in January 1871, and then he headed for the last place where Livingstone's presence had been reported—Lake Tanganyika, which is eighteen miles long and forty-four miles wide. In the twenty-first century, it is a part of four nations—Burundi, Zambia, Tanzania, and the Democratic Republic of the Congo. Despite the vague and outdated nature of the information available to Stanley, he came face-to-face with the ill Livingstone in November 1871 and famously greeted him by saying, "Dr. Livingstone, I presume?"

Stanley's journal served as the primary source for the documentary. Almost a century of change made it impossible for Glenn to follow the exact trails that Stanley had used; still, the production team planned to have him duplicate Stanley's general route. The trek would consume about six weeks. Glenn took his son, Dave, a Harvard student, along on the expedition. The twentieth-century explorer had been asked not to research Stanley's exploits; Wolper aimed to capture his genuine emotions throughout the trip.[72]

Glenn told viewers, "I come to Africa for the first time in an ancient dhow, the way Stanley did."[73] He and the TV crew sailed out into the

John Glenn on a boat in Africa during filming of the *Great Explorations* documentary. *Courtesy of the Ohio Congressional Archives*

Indian Ocean and landed on the beach in Dar es Salaam, Tanzania. After a few days of planning, they set out walking for a couple of days and later used Land Rovers, trucks, and even a steam-powered train, not one of which was available to Stanley. Despite the transportation advantages, the party often spent its nights in places where Stanley had camped. In one spot, villagers asked whether Glenn had a hunting license because a rogue elephant had been eating their corn crop. The village had formally requested that a game warden kill the animal; nonetheless, no warden had yet arrived. After driving around the area for a while, Glenn spotted the one-tusked elephant with binoculars. "Then I got out and went downwind and worked my way up very carefully hiding behind one termite mound to another and behind the bushes . . . until I got up close. And the elephant apparently it sensed I was there and charged. It was like something you would see in the movies," said Glenn, who already lived a life like something out of the cinema. He shot the elephant once, chased it, and struck it again. Villagers butchered the dead creature and dried its flesh like beef jerky.[74]

Dave Glenn kept a journal of the expedition. Perhaps realizing that, because of his father's fame, his journal might someday be found in a library, Dave referred to his father throughout as "Glenn."[75]

Wolper sold the show to ABC as a one-hour documentary, set to air January 11, 1968. The producer subsequently lost his sponsor before he could proceed with episode two, about Genghis Khan.[76]

For Glenn, there would be no TV series, but the year 1968 would carve a jagged path through American history that would feel as unreal as any television production. For Glenn personally, devotion to a friend would engulf him in new ideas and fresh experiences. A year of triumph and tribulations would change his life and set a new course for his future.

9

GLENN'S FRIEND, ROBERT KENNEDY

In the early morning hours of June 5, 1968, Senator Robert F. Kennedy gave a thumbs-up, saying, "And now it's on to Chicago, and let's win there." In a familiar mannerism, the New York Democrat reached up and brushed an errant lock of longish hair away from his blue eyes. Turning away from the crowd of supporters, he walked toward a back exit from the Ambassador Hotel ballroom in Los Angeles. Thus ended the last public statement of his life.

As friends and bodyguards Roosevelt "Rosey" Grier and Rafer Johnson led him through a kitchen pantry, a slight young man closed to within a foot of Kennedy. He was holding a campaign poster with a .22 pistol hidden inside. The dark-haired gunman fired multiple times, hitting Kennedy in the head and body. The gunfire also injured several bystanders. TV cameras broadcast images of Kennedy lying on the floor in a pool of blood with his eyes open and an apparent look of resignation on his face.

Pandemonium swept through the ballroom. Upon hearing the loud bangs, Kennedy's supporters instinctively knew what had happened. Like his brother less than five years before, Robert Kennedy had been gunned down. Shrieks and tears emerged from the frightened crowd. Kennedy supporters pressed toward the exits. It was unbelievable— and yet in 1968, it almost seemed inevitable, coming just two months after the slaying of Martin Luther King Jr.

Upstairs at the hotel, John and Annie Glenn watched the horrific scene unfold on television. Glenn raced down to the ballroom, but the mayhem blocked his way. He was too late to see Kennedy, a close friend, before he had been loaded into an ambulance. The Glenns rushed to the hospital. After three hours of surgery that left RFK comatose, Kennedy's pregnant wife, Ethel, who had ten children already, decided to do something about five of her children who were asleep at the Beverly Hills Hotel. To spare them from an all-too-public deathwatch, she decided they should return to their Hickory Hill home in McLean, Virginia. She chose John and Annie Glenn as the best people to care for the children, escort them home, and tell them that their father had been shot. Boarding an air force jet sent by President Lyndon Johnson, the Glenns returned the children to their home and spent the night with the young Kennedys. The following morning, a devastated John Glenn joined Susie Markham, a Hickory Hill neighbor, in an incredibly difficult task. They sat on the bed of each of the Kennedy children and revealed the terrible news: their father had died overnight in Los Angeles.[1]

For John Glenn, the turbulent 1968 established the trajectory that would guide the rest of his life. His involvement in Kennedy's campaign and his role in events following the assassination drew him into the world of politics, renewing his impetus to serve the public. He would never lose the blood-soiled memories of Los Angeles, but by year's end, he would start assembling the pieces that would complete the puzzle that was his future.

Glenn's years of friendship with Robert Kennedy were short, but for the former astronaut, this relationship became a life-changing experience. The two had met when Glenn was an astronaut visiting President

Kennedy in the White House. He went there February 5, 1962, just fifteen days before his historic orbital mission, to explain to a curious president exactly what NASA hoped to accomplish by placing him in orbit. The president asked whether Glenn believed everything possible had been done to make his flight safe, and Glenn said he did. Kennedy also questioned Glenn about what he expected to feel on liftoff.[2]

Not long after the flight of *Friendship 7*, President Kennedy invited Glenn and his family to join the Kennedy family at the presidential retreat, Camp David. That's where Glenn first understood the true meaning of "touch football" in the Kennedy family lexicon. At one point in a rough game, the astronaut tumbled into a woodsy patch of ground. Robert Kennedy darted over to check on him and began speculating aloud about headlines announcing that Glenn had survived the first US orbital flight only to suffer serious injury in a Kennedy family touch football game.[3]

Over time, Glenn and the attorney general became closer. "While I liked President Kennedy and felt comfortable with him," Glenn wrote in his memoir, "I found a deeper level of compatibility with Bobby and his wife, Ethel."[4] The friendship grew after JFK's assassination led to RFK's departure from the Johnson administration and election to the US Senate from New York. The Glenns regularly attended parties at Hickory Hill. While Glenn had grown comfortable with the Kennedys, he was more than a bit intimidated by the Hickory Hill dinner crowd. "You go there for a small dinner party," Glenn said, "and you may—well, you'll have a Supreme Court justice, and maybe a student militant from one of the universities, and maybe a civil rights leader from the South, and somebody who just happened to be in town that they knew from some time back that wasn't well known, and this is what made things so interesting."[5] Glenn's friendship with Kennedy broadened his sphere of acquaintances in public life.

The two families vacationed together, floating down rivers, waterskiing, sailing, swimming, and skiing. Lying on a beach soaking up sun was not a vacation for the active Kennedy. He and his wife liked being outdoors, careening through activities, taking risks, and challenging their children.

Competitiveness was a quality that had been nurtured in the Kennedy family since the childhoods of JFK and RFK. In early 1968, newspapers reported that Bobby's Bombers had squared off against Kennedy brother-in-law Stephen Smith's Torpedoes in a giant slalom ski race on the slopes of Waterville Valley, New Hampshire. The Torpedoes won, much to the senator's dismay. Participants on the two sides included Robert and Ethel Kennedy plus seven of their children; Eunice Kennedy Shriver and three of her children; Patricia Kennedy Lawford and her four children; Steven Smith, his wife Jean Kennedy Smith, and two of their children; singer Andy Williams and his wife, Claudine Longet; as well as John and Annie Glenn and their son, Dave.[6]

The Kennedys liked to win, and they attracted friends who shared their hunger for victory. Once, on a float down the Salmon River in Idaho, Glenn and mountaineer Jim Whittaker complained, "Bob was the only one who never overturned into the icy water." Glenn and Whittaker accused Kennedy of cheating by secretly placing rocks in the bottom of his kayak for added stability. "Our egos were rather severely bruised," Glenn admitted.[7] Often, these trips drew media attention. Photographers seemed ever present on the periphery, snapping photos of Glenn with the Kennedys.

Glenn admired many facets of Robert Kennedy. He enjoyed witnessing Kennedy's affection for children in general and the way he found time even on busy days to touch his own children—Kathleen, Joseph II, Robert Jr., David, Courtney, Michael, Kerry, Christopher, Max, and Douglas.[8] Kathleen was the eldest at fifteen in 1968, and Douglas, the youngest, was a little over a year old. No matter how frenetic the demands on Kennedy's time, "there was always time to hear a child's problem, give it serious thought and come to a mutually arrived at decision," Glenn observed. "There was a time to play, a time to share, and a time to be stern, for he was a father in the finest sense of the word." Glenn recalled that Kennedy "encouraged the best in them. . . . The family was never far from thought."[9]

Kennedy always encouraged his children to face their fears. After all, as a child, the would-be senator had felt an urgent need to learn how to swim, so he had thrown himself off a boat and landed in

Nantucket Sound, where he would either learn to swim or die at a very young age. "It showed either a lot of guts or no sense at all, depending on how you looked at it," JFK later said. Oldest brother Joe pulled young Bobby from the water.[10]

During one of their shared float trips, Glenn watched as the older Kennedy children dived from a ledge into the waters of Idaho's Salmon River. Afterward, each of the younger children cautiously edged toward the rim of the ledge and considered whether to take the plunge. The wheels were turning in their heads as they tried to conjure the courage to make a leap. With a combination of "coaxing, pleading, cajoling, and mocking," RFK convinced each of his children on the trip to either dive or jump from the ledge, and he took time later in the evening to congratulate each one and let them know that he was proud to have such children.[11] Diving or jumping into the water wasn't the point; daring to do it despite natural fears was what Kennedy wanted to see from his children.

At the same time, Glenn believed Kennedy had a deep effect on his own two children, who were in their late teens and early twenties during their father's friendship with Robert Kennedy. He remembered that on a flight from New York to Hyannis Port for a weekend break, Kennedy had mesmerized them in a discussion of the disadvantages faced by Puerto Rican children in New York City. Glenn's children listened as Kennedy spoke with passion. "As he talked, he would pound one hand into the other, stand up, sit down, would lean back in his seat, and talked with the utmost intensity regarding this intolerable situation that must be improved."[12] And when RFK spoke, the Glenn children listened.

Insatiable curiosity, "which led him to seek the opinions and ideas of some of the finest thinkers in our land," was another part of Kennedy's personality that Glenn found compelling. He noticed how deftly Kennedy could absorb new ideas, mesh them with his own perceptions, and transform them into workable plans. Glenn called this ability "a rare talent, as we can better appreciate by comparison to some of his contemporaries."[13] Glenn himself thought curiosity was a quintessential American quality. One Glenn document on "The

Age of Imagination and Inquiry" encourages Americans to maintain the curiosity of a child—something Kennedy had done.[14] Although Glenn's document is undated, it is written on NASA stationery, so it was written no later than 1964, which was early in his relationship with Robert Kennedy. A shared feeling about the importance of curiosity may have contributed to their friendship.

Even in the realm of personal lives, Kennedy seemed to be on an unending quest for insight. He had a habit of asking a roomful of friends a deeply personal question in hopes that their answers would broaden his own understanding of humanity. Once, as a group gathered in the warmth of a campfire, Kennedy questioned Glenn about his choice of the name *Friendship 7* for his Mercury capsule. Glenn explained that his children had selected the name from a list of possibilities, but that is not where the conversation ended. Kennedy carried his listeners in an unexpected direction with the goal of exploring the true meaning of friendship. He asked his fellow risk-takers what made a friendship "real" and how many true friends each one had. And by "true friends," he meant people you could trust completely. "Some had trouble getting beyond the fingers of one hand," Glenn recalled, and after RFK's assassination, he added, "I can now count one less."[15]

At a New York dinner party, Kennedy asked, "Outside of proposing to your wife, what was the most important event or decision of your life that had the greatest effect?" Among the guests who faced this inquiry were actor Sidney Poitier, director Mike Nichols, actor Rod Steiger, and Glenn. The question was personal and so were the answers. Surprisingly, Steiger cited his decision to enter a civil service job after leaving the military. Looking back, he could see that the job indirectly led to his involvement in a community theatre group, which set into motion his career as an actor. Glenn told the group that his biggest turning point was the decision to remain a marine after World War II. That choice had opened the door to work as a test pilot and later as an astronaut. World War II and its end likely became pivotal moments in guiding the lives of many in Glenn's age cohort. "We were all very much impressed with the fact of how intertwined the lives of many people are," Glenn said, "for just a slightly changed decision as a result of a few chance remarks or meeting particular people at a certain time would have changed the pattern of each of our lives tremendously."[16]

Kennedy, in Glenn's view, made the most of his financial wealth, using it as "a platform from which to rise, to act, to right injustice he saw in our midst." In Kennedy, Glenn saw great empathy for those less fortunate than he was. Often, a privileged childhood leads to a life of leisure and indifference to the problems of others. Glenn saw the opposite in his friend. Kennedy repeatedly stated that being satisfied with a nation in which the American dream was not accessible to every person was "just not acceptable."[17] He had higher goals for the United States.

"All his efforts were dedicated to an 'appeal to a tardy justice,' for he felt that every man's voice must be heard in some council of government," wrote Glenn. "Instead of despair, he saw hope that this country can be even greater when the injustices and inequities of opportunity for all are corrected." After RFK's death, Glenn found resonance in the eulogy delivered on the Senate floor by New York's other senator, Jacob Javits. In it, the liberal Republican said, "He had, so far as I know, the deepest concern for the underdog of anyone I had ever met." Javits especially noted his attachment to those who were "disenfranchised in terms of opportunity and in terms of the legacy to which all Americans are entitled. He was not the only man in public life to have this feeling in his heart; but, in my judgment, it burned in him more brightly than in any other man I have ever known."[18] Glenn believed Kennedy "felt that he had a responsibility to help improve other people's lot."[19] Although Glenn was a few years older than Kennedy, it is obvious that RFK became a political role model for him.

The Glenns also came to know the lighter side of Senator Robert Kennedy. Once, when they had spent the night at the Kennedys' apartment in New York City, they were slipping out early in the morning and had carried most of their bags to the elevator when Kennedy stuck his head out the door to wish them well. He realized that one of their bags still stood at the apartment door. After looking up and down the hall and seeing no one else, he ran out into the hall wearing a shirt and tie and underwear with bare feet and legs. He deposited the bag beside them and scurried back into his apartment before anyone else could see him.[20] With no paparazzi to capture his antics, he felt free to be mischievous.

Robert Kennedy lived his life as if he expected it to be short. Glenn noticed how that played out in daily life: "If you're going to play touch football, then let's play it hard. And if you're going down the river, why, let's go through the rapids. And if you're going to climb a hill someplace, well, let's go really all out to get to the top." One man cannot experience everything, even in a long lifetime. Therefore, Kennedy questioned adventurers who had achieved feats he would never experience, such as flying in space or climbing Mount Everest, and Glenn believed that in this way he enjoyed a vicarious experience that was important to him.[21]

In 1968 Kennedy struggled with the decision of whether he should run for president. He opposed the continuing war in Vietnam,[22] and in that year, American attitudes toward the conflict were especially volatile. The incumbent president, Johnson, was a Democrat like Kennedy, and that made the senator hesitate, thinking that perhaps a campaign against a sitting president would be a fool's errand. He and Johnson had an especially difficult, intensely bitter relationship. They never liked each other, and things worsened after LBJ became president following the murder of Kennedy's brother. RFK hung back in 1968 until Minnesota senator Eugene McCarthy, an antiwar candidate, made a respectable showing against Johnson in New Hampshire's Democratic primary. On March 16, Kennedy jumped into the race. Fifteen days later, an emotionally bruised and battered Johnson withdrew his name from consideration and promised to focus his attention on Vietnam peace talks.

Kennedy's campaign push began with disappointment. Many public figures labeled him as an opportunist, and there were personal letdowns too. He told Glenn that one of the things he had learned in politics was that "friendship" was a negotiable arrangement rather than a true and loyal relationship. He felt that many people who had been his friends, even some who had encouraged him to run for president, turned their backs on him when he announced his candidacy. "There were a couple of the people in Washington who he thought were very close friends of his who did not support him in his candidacy," Glenn said. "This disappointed him very much."[23]

Kennedy had experienced something similar when he lost influence after Johnson replaced his brother as president.[24] RFK once said that the most you could hope to find in politics was "an arrangement of alliances" keyed to a particular time and a particular situation.[25] (Glenn discovered the same phenomenon in his race for an Ohio Senate seat in 1970; even Kennedy family members failed to support his primary battle.)

As counsel in congressional investigations and as attorney general, Kennedy had a reputation for being "ruthless," but the man Glenn knew was shy and sensitive.[26] Some observers believed that his brother's assassination softened Bobby Kennedy; some thought his reputation for ruthlessness sprang from political enemies' exaggerations; others saw the fire of a ruthless prosecutor burning still within the passionately liberal senator.

In his first solid commitment to politics since his aborted senatorial candidacy in 1964, Glenn and his whole family jumped into Kennedy's presidential campaign wholeheartedly. Glenn notified Jim Webb at NASA about his decision, and Webb tried to discourage his involvement. Nevertheless, Glenn remained true to his friend.[27] Often, Glenn traveled with Kennedy, but sometimes he campaigned alone in rural areas where only a tiny audience waited to hear his message. Kennedy chose to focus his attention on primaries in Indiana, Nebraska, Oregon, South Dakota, and California, with California clearly being the most valuable prize. In the opening days of the campaign, Glenn accompanied the candidate to the University of Alabama in Tuscaloosa. A southern university appeared to be an unlikely spot for an outspoken civil rights advocate to begin campaigning in 1968, but Kennedy voiced high praise for the university's president, Frank Rose, whom he credited with establishing an atmosphere open to hearing a variety of viewpoints. He joked that there was bad news: legendary University of Alabama football coach Bear Bryant "will not run as my vice president." The students cheered Kennedy and were delighted by the surprise guest on the platform—an astronaut! Glenn told them that he supported Kennedy "because he tells it like it is."[28] He believed that RFK chose a southern swing early in the campaign to counter political foes who had demonized him in that region, condemning his civil rights stands as attorney general and as a senator. And after that visit to Dixieland, Glenn concluded, "He got a far better reception there than he anticipated or than I certainly had anticipated."[29]

While Kennedy may have won at least a few southern fans, Glenn was losing some. He ran into opposition at Georgia-born Royal Crown Cola, where he served on the board and worked. Some of his colleagues were conservative southerners, and they disapproved of a company board member playing such an active role in a liberal candidate's presidential campaign. He heard that they planned to introduce a resolution banning board members' active participation in politics. His response was simple: if they did that, he would resign and make a public statement that cast Royal Crown in a bad light. The resolution died there.[30]

Glenn was not with Kennedy in Indianapolis on the night Martin Luther King Jr. was killed. When Kennedy heard the news of King's death, he was with a *Newsweek* reporter, who described how Kennedy "seemed to shrink back as though struck physically." He raised his hands to his face and said, "Oh, God. When is the violence going to stop?"[31] Despite warnings from local law enforcement officers, Kennedy faced black audience members who were unaware of the assassination, and in a brilliantly improvised speech, he delivered the shocking news and managed to make a strong case for a peaceful response to this tragedy. Glenn read the speech, which he called "an eloquent appeal to the people to remain calm and not lose their heads, and not go off and do a lot of things in retaliation that we would regret later." He remembered Kennedy's fears about the kind of violent response that occurred over the following days.[32] Riots erupted in more than one hundred American cities; thirty-nine people died, and more than twenty-five hundred suffered injuries.[33] The National Guard patrolled streets in many cities. The deaths, the flames, and the marching soldiers combined to create a scene tailor-made for a dystopian novel.

Glenn campaigns alongside Robert F. Kennedy during Kennedy's 1968 presidential campaign. ©*Burton Berinsky, by permission*

Although no direct connection can be drawn to Kennedy's speech, Indianapolis interestingly was one American city that did not experience violence.[34]

Later in April, a spur-of-the-moment campaign decision sent Glenn home to Ohio, where he and former assistant secretary of state Roger Hilsman led a rally for Kennedy at Ohio State University.

Glenn told the students that their generation was undergoing a crisis of confidence as "the first generation to grow up with the scientific method and they are extending it to government, philosophy and society. They want to know why." He urged students to get involved in the campaign, with special attention to canvassing for Kennedy in preparation for the Indiana primary. That plea was the underlying reason for Glenn's sudden assignment in Columbus.[35] In May, Glenn hit the road on his own, visiting four Nebraska cities, where he held press conferences and delivered speeches,[36] and on June 3 he campaigned solo again, making several stops in South Dakota.[37]

In light of Nebraska's farm-state status, Glenn enjoyed watching Kennedy make light of his background in agricultural affairs. He often claimed that he, his wife, and their ten children were consuming more agricultural products "than any family he knew." At other times, he told crowds that the state of New York "leads the world in the production of sour cherries."[38]

Kennedy won in Indiana and Nebraska but lost to McCarthy in Oregon, making McCarthy the first candidate to defeat one of the Kennedy brothers.[39] This loss heightened the importance of a win in California, where the primary was scheduled for June 4, the same day as the South Dakota primary. Kennedy had been spending time in California whenever he could. He did whistle-stop tours, and when he was not meeting with his many high-level supporters in the Democratic Party, he could often be found walking the streets of poor neighborhoods. Glenn recalled that RFK's favorite moments were conversations with youngsters. "The children ran, jumped, laughed, and sought his hand," Glenn wrote. "He in return talked and joked with them as we walked through the midst of squalor—unpainted homes and unpaved streets."[40]

One of the most dramatic events in Kennedy's California campaign was a meeting after midnight in West Oakland. Kennedy had agreed to talk secretly with many of the Bay Area's most prominent black leaders, including some radicals. Glenn planned to go with him, but Kennedy warned, "This may not be a pleasant experience. These people have got a lot of hostility and lots of reasons for it. When they get somebody like me, they're going to take it out on me. . . . But no matter how insulting a few of them may be, they're trying to communicate what's inside them."[41]

Glenn went along anyway. Inside a church, Kennedy met with about two hundred African American leaders—and his hosts initially insisted that he be the only person from the campaign admitted to the gathering. When the leaders learned that one of the people waiting outside was John Glenn, they let him join them. Later, they allowed Olympic star Rafer Johnson to enter as well. According to Glenn, "It was the whole spectrum of Negro leadership." The attendees ranged from professional leaders of the African American community to Black Panthers. In the two-hour meeting, those gathered told Kennedy about rumors that the old Japanese internment camps would be reopened to house African American leaders on the West Coast. Some people launched verbal attacks on Kennedy, but RFK told Glenn that he understood many of the harshest accusations against him were made to impress the other attendees rather than to attack his credibility or his record.[42] "Many of our national leaders would not have been physically safe in that area and meeting, but Bob came to them, learned from them, and they from him, with mutual understanding and dignity," Glenn remembered.[43]

As Kennedy and McCarthy battled it out in primary states, there was a likely spoiler waiting in the wings to claim victory without competing in any primaries. Vice President Hubert H. Humphrey was accumulating much of the old party support that would have gone to Johnson if he had stayed in the race. Humphrey was an ebullient and sincere traditional liberal. As a young mayor of Minneapolis, he had waved the civil rights banner in 1948, long before national politicians would even consider supporting equal rights for minorities. He had a delicate relationship with Johnson, who sometimes treated him as an ally and at other times acted as if Humphrey were a mindless sycophant.

Humphrey's big decision in 1968 was where he should stand on the Vietnam War. While Johnson was hugely unpopular in many portions of the party, he still had the power to make or break Humphrey, so openly opposing Johnson's Vietnam policy in the months leading up to the Democratic National Convention would have been political suicide for Humphrey. With the political machine backing Humphrey, the best Kennedy or McCarthy could hope to achieve was an honest chance to compete at the Democratic National Convention in Chicago in late August.

During the campaign, when Kennedy hit a bump in the road—and Humphrey's candidacy was more like a hefty boulder—he sometimes jokingly blamed Glenn's daughter, Lyn, who had encouraged him to commit to the race before he had made his final decision. Lyn was with her parents and Kennedy one day when they were campaigning in San Francisco. Kennedy, along with John, Annie, and Lyn Glenn, jumped on a cable car headed downtown. A crowd jumped aboard with them, and others gathered on the sidewalks. In Glenn's words, "People were just hanging all over the car." Lyn, who was standing near the driver, asked him how he could ever hope to collect fares from all those people, and he said he wouldn't try with Kennedy aboard. He said he couldn't afford to help RFK's campaign, but "the least I can do is ring my bell for him anyway."[44] And he rang his bell.

Many days on the campaign trail were stressful. One thing the dog-loving Kennedy did to make his days lighter was traveling with his springer spaniel, Freckles. In motorcades, Kennedy and Glenn would sit on the back of a convertible and wave to the crowd. Often before the cars had reached their final destination, Freckles would climb up from the back seat and sit between them so that the two men looked like the canine candidate's entourage. As Glenn remembered it later, Kennedy said that "the day that Freckles started shaking hands with people, that was the day he would quit."[45]

Although McCarthy bragged about the army of college students who cut their long hair and shaved their beards to "get clean for Gene" and bolster his campaign, Kennedy tended to be a hit when speaking to college campuses. "Probably one of his strongest audiences was always a student audience because they seemed to identify with him," Glenn said. "And he was able to talk to them and share views with them that other politicians could not do."[46] Kennedy typically ended each appearance with an open discussion to create a dialogue with the audience. "I think he enjoyed the question-and-answer periods more than really giving a set speech because it was not only a challenge but it's where he also got a feel for what the people were really concerned about, what they were thinking about," Glenn said.[47] The former astronaut always spotted some McCarthy signs in the audiences, and Kennedy had a way of turning those placards into a joke. He would say "that he was sorry that fellow hadn't learned to spell Kennedy, yet." Then, he would spell: K-E-N-N-E-D-Y.[48]

Mississippi newspaper editor and publisher Hodding Carter Jr. dined with Glenn, RFK, Rosey Grier, and Rafer Johnson on the night before the California primary, and the mood was ebullient. Carter said of Kennedy, "I never saw a man as happy in my life. . . . We knew it was in the bag, and it *was* in the bag."[49] For Kennedy, attaining victory in California and South Dakota on the same day was a great way to close the primary season with triumphs in a highly urbanized state and a mostly rural state at the same time. Indeed, June 4 was a great day for Kennedy at the polls. The celebration at the Ambassador was rousing, especially after Kennedy's loss in Oregon.

Then Robert Kennedy turned and walked into the space where Sirhan Sirhan awaited. Sirhan was a Palestinian Jordanian in his early twenties who had lived in the United States since he was twelve years old. He fired at Kennedy before being disarmed and detained by Grier, Johnson, and author George Plimpton, a Kennedy friend. Authorities later determined that Sirhan's motive was opposition to RFK's support for Israel.

Though Sirhan was captured by witnesses who saw him fire the gun, conspiracy theorists have alleged that he had coconspirators, most notably a young woman in a polka-dot dress seen by several witnesses. Others believe that a second gunman was in the pantry or that Sirhan had been "programmed" to shoot Kennedy and become the fall guy for the assassination. Conspiracy theories, like assassins, seem inseparable from the stories of John and Robert Kennedy. Sirhan accepted responsibility for the crime in 1969 and received a death sentence. Later, California outlawed the death penalty. Sirhan, now an elderly man, continues to serve a life sentence more than fifty years after the assassination.

From the moment Sirhan fired, Glenn's life changed dramatically. He had lost a friend and a leader whom he admired greatly, but Glenn never believed in skidding to a stop in the face of adversity. On the day of RFK's death, he could be seen on the seven-acre Hickory Hill estate playing touch football with Kennedy's sons, David, twelve, and Michael, ten, while the toddler, Douglas, watched. Others of the children played stickball and swam.[50] Syndicated newspaper columnist Phyllis Battelle wrote, "No man, not even a perfectionist and devoted father like Robert F. Kennedy, could command a finer comrade for his sons."[51]

When Glenn found a few quiet moments, he wandered into Kennedy's study and paged through books out on the desk. Kennedy made lots of marks in his books, indicating passages he liked or found thought-provoking. Among the books, Glenn picked up a collection of poems and essays by Ralph Waldo Emerson. RFK had marked three passages from Emerson, and Glenn saw a lot of his murdered friend in them:

If there is any period one would desire to be born in, is it not the age of revolution, when the old and the new stand side by side, and admit of being compared; when the energies of all men are searched by fear and by hope; when the historic glories of the old can be compensated by the rich possibilities of the new era? This time, like all times, is a very good one, if we but know what to do with it.

What is the remedy? They did not yet see, and thousands of young men as hopeful now crowding to the barriers for the career do not yet see, that if the single man plant himself on his instincts, and there abide, the huge world will come round to him.

The characteristic of genuine heroism is its persistency. All men have wandering impulses, fits and starts of generosity. But when you have resolved to be great, abide by it, and do not weakly try to reconcile yourself with the world. The heroic cannot be the common nor the common the heroic. Yet we have the weakness to expect the sympathy of people in those actions whose excellence is that they outrun sympathy and appeal to a tardy justice. If you would serve your brother, because it is fit for you to serve him, do not take back your words when you find that prudent people do not commend you. Adhere to your own act and congratulate yourself if you have done something strange and extravagant and broken the monotony of a decorous age. It was a high counsel that I once heard given to a young person, "Always do what you are afraid to do."[52]

Parenting Kennedy's children, sitting in his study, and reading the notes he had scrawled in the margins of his favorite books, Glenn inhabited a strange position in the twilight world between the moment when you learn of a loved one's death and the moment when it becomes real to you as you begin to understand how much that person's absence will change your day-to-day existence.

The Kennedy family planned an elaborate, unforgettable farewell to the late senator, a parting that was new and yet—as Americans saw the faces foremost among the mourners—hauntingly familiar. Even Caroline Kennedy, now ten, and John F. Kennedy Jr., seven, represented echoes of an earlier funeral in which they played a prominent role; now almost five years older than they had been at their father's death, they seemed battle-weary. Many Americans consoled themselves in 1963 that the Kennedy children were too young to fully understand what had happened; in 1968, they knew.

Someone among the Kennedys' friends and staffers prepared a distribution of responsibilities for those close to the family. The duties included funeral theme, transportation, New York coordinator, New York advance, Washington coordinators, Washington advance, security, invitations, printed materials, Kennedy family coordinator, and finally, RFK family control officer—a job assigned to John Glenn.[53] The former astronaut also would serve as a pallbearer on the right side of the casket. Every detail of who would stand in what spot had been worked out in advance. The ten other pallbearers would be RFK's brother-in-law Stephen Smith, JFK's prep school roommate and Kennedy family friend Lemoyne Billings, former RFK aide and newspaper editor John Siegenthaler, former British ambassador to the United States David Ormsby-Gore, former secretary of defense Robert McNamara, former ambassador-at-large Averill Harriman, retired general Maxwell Taylor, mountaineer James Whittaker, former secretary of the treasury Douglas Dillon, and RFK's friend David Hackett. Senator Edward "Ted" Kennedy, the last surviving Kennedy brother, and Joseph P. Kennedy II, RFK's oldest son, would walk at the head of the casket.[54] Kennedy had named sons after more than one of the pallbearers.

The funereal rituals began with a Mass at Saint Patrick's Cathedral in Manhattan. To Glenn, the breadth of RFK's appeal was evident in the cathedral, where the last of the Kennedy brothers delivered a touching eulogy with a sometimes-quivering voice and where Andy Williams sang "The Battle Hymn of the Republic" like a celestial choirboy. "There were students, the poor, civil rights workers, city officials, high government officials, military, personal friends from boyhood days, members of the clergy of all denominations—all come to pay their respects," Glenn said. "The level of esteem with which he was held not only here, but around the world, was well-illustrated as I stood in the Honor Guard, felt a tap on my shoulders as someone prepared to take my place. As I stepped away, my place was taken by U Thant, secretary general of the United Nations."[55]

After the Mass, John and Annie Glenn joined roughly eleven hundred people on a train carrying the casket from New York City to Washington, where a procession would carry Kennedy's body to a burial plot near his brother's grave at Arlington National Cemetery. The eight-hour train ride turned the twenty-two cars into a moving community, where passengers shared memories of RFK. Some reporters even datelined their reports as coming from "ABOARD THE KENNEDY FUNERAL TRAIN." Intermittently, Ethel Kennedy; her oldest son, Joe; and Ted Kennedy walked through the cars comforting every friend and acquaintance, while others expressed wonder at the

Kennedys' sense of obligation to ease the pain of others after a death in their own family. The *New York Times*' Russell Baker wrote that within this community, "there must have been a poignant awareness that when the train reached Washington it was unlikely that they would ever all gather together again."[56] Outside the train, another story was unfolding. Americans of all kinds lined the tracks:

Five youngsters in swimwear, backs straight like soldiers at attention, standing alongside their parents.

A man leaning on crutches but waving one crutch as the train rolled by.

Some mourners holding US flags; others clutching wildflowers.

A couple holding aloft a hand-lettered sign that said SO-LONG BOBBY.

Women praying on their knees.

Five workers holding their hard hats over their hearts.

Young men, sometimes shirtless and dirty, saluting the passing train.

An honor guard displaying city, state, and national flags.

And thousands more.

In Elizabeth, New Jersey, two mourners were so enthralled by the Kennedy train that they failed to hear a northward-bound train, which struck and killed them. Some Kennedy friends witnessed the accident from the funeral train, adding to the horror of their week. From outside the train, the coffin could be seen, and at times, various Kennedys stood on the back platform to acknowledge those who had waited for hours on a steamy June day.

When the train reached Washington's Union Station at last, a motorcade left for the cemetery. Here, too, crowds lined the streets as drizzling rain fell. The automotive cortege drove slowly past the Capitol and the Justice Department, two places where Robert F. Kennedy had served his country. The cars also passed Resurrection City, a camp of huts and tents on the National Mall occupied by poor Americans protesting inequality. Some of the three to five thousand protesters camped there raised their fists in black power signs, while others sang "The Battle Hymn of the Republic."[57] At the cemetery, pallbearers delivered the casket to a resting place fewer than fifty feet from President Kennedy's grave. As a former marine, Glenn took charge of folding the American flag in military fashion and handing it to Ted Kennedy, who passed it along to his brother's pregnant widow.

When RFK's assassination so closely followed King's slaying, many shaken Americans embraced gun control as an important issue, and Glenn made a forceful step into politics by taking on the chairmanship of a new gun-control group known as the Emergency Committee for Gun Control. In this role, he promoted tighter laws on gun ownership across the country. Glenn had been given a head start; Johnson had initiated his own legislative attack on rampant misuse of guns by proposing legislation on registration of firearms and licensing of owners.[58] Glenn traveled the nation pushing for restrictions on gun ownership and urging Americans everywhere to write to Congress demanding tougher laws. When he appeared on New York City mayor John Lindsay's TV program, he acknowledged that he had six guns in his own home and that he enjoyed hunting. And yet, Glenn told Lindsay that he now felt gun restrictions made sense because "we're no longer a frontier prairieland: We're an urban society."[59]

Glenn undoubtedly knew that his call for restrictions on the sale of guns would generate criticism, particularly in rural areas of the country, including central Ohio, where he had grown up. Ken Gookins, who wrote the outdoor column for the *Times Recorder* in Zanesville, which is close to New Concord, defended Glenn's feelings of desperation in the wake of his friend's murder; however, he argued that Glenn should focus on capturing and punishing killers, not violating hunters' constitutional rights.[60] This was a popular theme for Gookins, filling at least three columns between August 1 and November 1 that year.

Gun control advocates often meet stiff resistance, but Glenn never shied away from his leadership role. He even achieved a measure of success. Legislation passed, although it was much weaker than Glenn had hoped. The White House invited Glenn to be present at the bill-signing ceremony.[61] On October 22, Johnson signed into law the Gun Control Act of 1968, which barred interstate sale of guns and ammunition, stopped the sale of lethal weapons to youths, and attempted to block the dumping of cheap foreign guns on the American market. It was not an extremely restrictive gun-control measure, but it may have saved some lives.

As 1968 neared its end, with Richard Nixon defeating Hubert Humphrey and George Wallace to win the presidency, many Ohio newspapers reported Glenn's renewed interest in running for office. A year filled with memorable politicking and deep grieving had expanded his sense of responsibility to the nation's citizens and his understanding of sophisticated campaigns. He would not be another Robert Kennedy; he lacked the fire and the drive to challenge injustices as vociferously as Kennedy had. That was not who John Glenn was. Nevertheless, as Glenn looked ahead to what he hoped would be a successful career in politics, RFK would remain a forceful figure in his thoughts. Glenn's Senate press secretary Dale Butland said that, even decades later, "when [Glenn] talked about Bob Kennedy, he would often tear up and become emotional."[62]

10

INTRODUCTION TO POLITICAL LIFE

For John Glenn, 1969 was a year to confront the empty promises of "might have beens." His friend Robert Kennedy might have been inaugurated president, taking the nation in an entirely different direction. And even more personally, John Glenn might have been the first astronaut to set foot on the moon. Instead, he watched from mission control as Neil Armstrong lived his former colleague's dream.

Of course, Glenn was forty-eight—ancient by early astronaut standards—when the lunar landing occurred on July 20, 1969. Nonetheless, as a Mercury astronaut, he had hoped to leave footprints on the dusty surface of Earth's only natural satellite. Of the Mercury Seven, only Alan Shepard would reach the moon. Shepard had been grounded throughout Project Gemini as a result of Ménière's disease, a sickness that causes fluid pressure to build up in the inner ear with resulting disorientation as well as nausea and vomiting. While he was off the flight list, he served as chief of the Astronaut Office. After five years, Shepard learned about surgery that could resolve the problem and rushed to have it done in 1968. He was reinstated and flew to the moon in the winter of 1971 aboard Apollo 14, at age forty-eight. He famously made two golf shots on the lunar surface.

For John Glenn, there was no chance. President Kennedy's request that NASA take him off the list of active pilots had sent his life in another direction. All he had been able to do was wistfully watch as the US space program marched toward the moon. He authored an article months before the landing, praising NASA's remarkably accurate navigation system that took Apollo 8 into lunar orbit in December 1968 and the engineering know-how that allowed Apollo 9 to successfully test the lunar module in Earth orbit in early 1969. "No single flight can be said to be more important than any other," he wrote, "for each flight builds on previous experience and each is on the cutting edge of science on its own day." Despite that principle, Glenn, like most Americans, seemed to find the moon-orbiting flight especially captivating. He believed there was a simple reason for the US space program's leadership in landing a man on the moon: "Our nation is in large measure what it is today because we have been a curious people, willing to back research and exploration, willing to inquire into the unknown, curious to push back the frontiers of the known, and put new knowledge to work in our free system."[1]

In another piece written in 1969, he contended that the Soviet Union's secrecy had enabled the United States to surge ahead in the rush to the moon. If the Soviets had been more open, he believed, the two programs could have helped each other. The United States and the Soviet Union had "different natural resources, different ways of thinking about science and space, and a different set of national priorities into which the exploration of space had to fit," he acknowledged.[2] A year later, he authored an article in the wake of Apollo 13's almost catastrophic flight to the moon. In it, Glenn argued that NASA had always known that astronauts' lives were in danger. He favored continuing moonshots, though he felt that in the near future, "we should put more emphasis on missions that contribute to solving problems on Earth."[3]

In 1970, Glenn started anew in his search for a place in public life. Campaigning for the US Senate from Ohio, Glenn once again faced a primary fight. He had stepped down from his executive position at Royal Crown Cola to devote his time to the race. Howard M. Metzenbaum opposed him in the contest to replace the retiring senator Stephen Young.

Metzenbaum was a Cleveland businessman and lawyer who carried Young's endorsement.[4] As chairman of a group of suburban weeklies, he led a media-savvy campaign. While lacking Glenn's fame, his rival had a more professional campaign staff and more money. Organized labor backed him, and he had built strong relationships with leaders of the state Democratic Party. He also had solid liberal credentials as the first non-Catholic to win the Interracial Justice Award of the Catholic Interracial Council of Cleveland. While Glenn defended high expenditures on the space program, Metzenbaum argued that more money should be spent to resolve problems on Earth, which he considered more vital than more moon missions "to bring back more rocks." While the businessman took a strong stand against the Vietnam War,[5] the former marine mostly questioned whether the war had been fought properly.[6]

When the two men faced off in an impromptu debate in Dayton on March 7, their biggest area of conflict was space program expenditures, with Metzenbaum accusing Glenn of shifting positions on the issue by saying that some reductions would be appropriate. Furthermore, he charged Glenn with retreating from his strong advocacy of gun control in 1968.[7] (Glenn remained a low-key gun control supporter throughout his career, supporting the Brady Handgun Violence Prevention Act in 1993 and the Federal Assault Weapons Ban in 1994.)[8]

Larry O'Brien, who served as JFK's legislative liaison and as Lyndon Johnson's postmaster general, became chair of the Democratic National Committee in March 1970 and was forced to teach Glenn the political facts of life. "John was very disappointed that he had not received strong support from the Kennedys in this quest," he said. "It surprised him; it disturbed him." O'Brien found Glenn's attitude naive, because high-profile politicians, such as Senator Edward Kennedy of Massachusetts, often considered it wise to stay out of intraparty contests. Glenn also had expected financial assistance from the Democratic National Committee, and O'Brien recalled, "I had to advise John there was no financial support; we had no resources for that."[9]

Early in the campaign, Glenn was the front-runner. As he hopscotched around the state, two Caddell Polls showed that Glenn's support dropped from 51 percent in January to 31.5 percent in March. Most of the support he lost moved into the undecided column rather than into Metzenbaum's camp, which indicates that Glenn backers changed their minds about the political novice without necessarily embracing Metzenbaum.[10]

The primary election occurred on May 5, 1970, one day after National Guardsmen opened fire on students at Kent State University in northeastern Ohio. Four students died in what became known as the Kent State Massacre. Some of the students were protesting US bombing of Cambodia as part of the Vietnam War; others were simply walking across the campus. This event suggested a continuing disintegration of order, and that horrified Glenn.[11]

When the primary votes were counted, Metzenbaum won by about 12,500 votes, thanks in part to a powerful TV advertising campaign that overwhelmed Glenn. Metzenbaum later lost the general election to Republican Robert Taft Jr., a member of the Taft political dynasty.

The Glenn family suffered more personal losses in early 1971, when John's mother and Annie's father died within ten weeks of each other.

In the aftermath of political defeat, Glenn set to work building the kind of support he would need to capture a Senate seat in the future. He studied his loss and drew political lessons about the need to build stronger ties within the party leadership. He nurtured connections with Democrats around the state and tried to maintain a high profile. Riding in the pace car at the 1971 Indianapolis 500 was one means of remaining in the public eye. Unfortunately, the stunt came dangerously close to killing the former astronaut and wannabe senator. A car dealer was driving the Dodge Challenger when he lost control while traveling at 125 miles per hour and smashed into a stand where photographers were positioned. No one was killed, but at least thirty photographers were injured, two of them seriously. The spectacular crash miraculously left the car's passengers with only minor injuries.[12]

While John Glenn awaited another political opportunity, Annie Glenn reached a turning point in 1973. After a lifetime of struggling with severe stuttering, she entered a new three-week program at Hollins College in Roanoke, Virginia. While there, she labored over exercises eleven hours each day and embraced a new approach to speaking. Exercises focused on alphabetical sounds in individual syllables, and by the end of the program, she was able to complete multiple sentences without stuttering. Though she didn't consider herself "cured," this represented a huge leap after a lifetime of relying on intermediaries to communicate for her in many situations. At the end of her studies, she called her husband, and he cried to hear her speaking in complete sentences.[13]

Later, John watched tearfully as Annie read to their first grandchild—something she never had been able to do with her own children. "Annie inspired a lot of people to get treatment for their disabilities," Glenn wrote proudly. "Through the American Speech and Hearing Association (ASHA), she would meet with people to tell her story and how she stuck with her treatment. ASHA was impressed enough by Annie's accomplishments that the organization now grants an award in her name. The 'Annie' honors people who overcome great communication difficulties and achieve distinction by helping others."[14] For years, she made a point of attending every ASHA convention "because communication and what our ASHA members do for others is so important to her," said Dawn Valasquez, ASHA's associate director of accreditation for research and quality management. "She just represents the ability to overcome, to say that I have a challenge," said Tess Kirsch, the organization's associate director of accreditation for policy and education.[15]

Diminishing her stuttering meant that Annie was able to participate more fully in group activities. She served on the National

Deafness and Other Communication Disorders Advisory Council of the National Institutes of Health and on the advisory panel of the Central Ohio Speech and Hearing Association.

Meanwhile, accusations of obstruction of justice against Richard Nixon in the Watergate scandal had unexpected repercussions in Ohio politics. Nixon, under siege by a special prosecutor as well as congressional investigations, ordered Attorney General Elliot Richardson to fire Special Prosecutor Archibald Cox. Richardson resigned in protest, and Deputy Attorney General William Ruckelshaus followed suit. After being called to the White House, Solicitor General Robert Bork dismissed Cox. The so-called Saturday Night Massacre on October 20, 1973, led to the January 1974 appointment of Republican senator William Saxbe of Ohio as attorney general.

As Nixon struggled futilely to hold on to the presidency, Ohio governor John Gilligan, a Democrat, had an opportunity to name a Democrat to take Saxbe's place. Glenn reports in his memoir that Gilligan wanted him to run for lieutenant governor in 1974 on the governor's ticket, strengthening Gilligan's position as a possible candidate in 1976 for the presidency or vice presidency. Glenn was not interested. He believed that Metzenbaum promised to support Gilligan's reelection campaign in return for the interim appointment to the Senate, so Gilligan named him to fill Saxbe's seat. Glenn was unhappy with the choice, but he was prepared to face Metzenbaum again in the Democratic primary in May 1974.[16]

This time, Glenn's campaign had stronger organization, and a well-thought-out strategy. He had won seventy-six of eighty-eight counties in 1970, but Metzenbaum had triumphed by carrying many of the most populous counties. Glenn's 1974 campaign began with a bus tour of the entire state, but after that, he focused on the heavily populated urban and suburban counties. [17] Metzenbaum still had the support of labor and invited backers to $100-a-plate dinners, while Glenn staged a poor man's campaign. On April 19, the *Plain Dealer* in Cleveland paired two articles: one focused on Metzenbaum's black-tie crowd in a hotel ballroom raising $85,000,[18] while the other described Glenn standing on a table to address five hundred Ohioans who dined

John and Annie Glenn take a moment to relax during the 1974 Senate campaign. *Courtesy of the Ohio Congressional Archives*

on corned beef and cabbage for ninety-nine cents a plate at a café.[19] The contrast worked in Glenn's favor.

In addition, Glenn won support by drawing attention to Metzenbaum's tax returns. Glenn released his returns on February 18. He had a net worth of $767,800 as a result of salary, savings, real estate holdings, partial ownership in several Holiday Inns, and other investments. Glenn paid taxes every year.[20] At that point, Metzenbaum had "released no income tax returns, declined to list his income or taxes paid by years and refused to itemize his $3.3 million in outstanding debts," according to a Citizens for John Glenn press release.[21] Two months later, when Metzenbaum released his tax returns, they demonstrated great familiarity with the intricacies of tax write-offs: he paid no income taxes whatsoever in 1969 due to losses in his business, and he had a $118,000 dispute with the Internal Revenue Service over deductions he had taken six and seven years earlier.[22]

Metzenbaum's failure to release the returns quickly and his incumbency led some voters to group him with Nixon and other Watergate wrongdoers.[23] Glenn did his best to reinforce that connection in voters' minds, saying Metzenbaum finally released his returns because "like President Nixon, he was desperate and forced to do so by public opinion."[24] Metzenbaum might have fared better if Glenn had received the interim seat and saved him from the stain of incumbency.

Glenn's opponent made another crucial error: in a Toledo union hall, Metzenbaum questioned Glenn's qualifications by asserting that Glenn had never held a job. Glenn waited two weeks until he made his final statement in their last debate, and his oratory soared like a rocket. In a strong speech, he talked about his twenty-three years in the Marine Corps and his service in two wars, and then he reviewed his service as an astronaut and said, "It wasn't my checkbook. It was my life that was on the line." Speaking sternly, he buried Metzenbaum in a patriotic message. "You go with me as I went the other day, out to a veterans' hospital, look those men with their mangled bodies in the eye and tell them they didn't hold a job. You go with me to any Gold Star Mother and you look her in the eye and tell her that her son did not hold a job. You go with me to the space program and you go as I have gone to the widows and orphans of Ed White and Gus Grissom and Roger Chaffee, and you look those kids in the eye and tell them their dad didn't hold a job."[25] Twenty-two seconds of uninterrupted applause followed—a lifetime in a campaign, said political analyst Mark Shields.[26] Metzenbaum later defended his position, saying he meant that Glenn had not had a career in private business—hardly a prerequisite for public office.

Glenn's still underfinanced campaign glided to victory over Metzenbaum with a margin of ninety-one thousand votes. He received strong support in urban areas—exactly what he had hoped to do. It had been an occasionally nasty campaign filled with charges and countercharges. Glenn's intemperate accusations were perhaps out of character given his nice-guy reputation, but Metzenbaum had set the tone. In the Watergate era, "the fact that he [Glenn] never saw a day in political office became a virtue," one writer concluded.[27]

Glenn trounced his Republican opponent, Cleveland mayor Ralph Perk, winning a 2–1 victory in November. He had begun his quest for

John Glenn delivers his powerful "Gold Star Mother" speech at a Cleveland City Club debate during the 1974 primary campaign.
Courtesy of the Ohio Congressional Archives

the Senate when he was just forty-three, and now he was preparing to take office at fifty-three. "There's still some of the Boy Scout in him," one of his campaign advisers said. "But this is a very refreshing guy to be around."[28]

The 1974 election brought in a new class of senators in a post-Watergate backlash vote against the Republican Party and incumbents in general. Glenn was a member of that class, and his arrival at this time in some ways shaped his start as a US senator. Being a Washington outsider had been a plus for Glenn and many other new arrivals, and that meant they were not restrained by the long-accepted political rules of the game.

Rather than sitting back and watching the Senate at work for a while before stepping into the fray, he dived into complex issues. "I think he realized that there would be naysayers, because he hadn't come the normal government route. . . . He just jumped right in sort of at the top," said David Hafemeister, who worked for Glenn as a

Science Congressional Fellow for eighteen months early in his term and returned to work for him in the 1990s.[29] Glenn later said that all of the technical details he juggled at NASA made him pursue difficult-to-unravel and sometimes flat-out dull issues, such as improving the government bureaucracy and establishing nuclear nonproliferation treaties—what one ex-aide called "not the big glam things."[30] Glenn and his staff called them MEGO (My Eyes Glaze Over) issues.[31]

In his early years on Capitol Hill, Glenn quickly developed a reputation for "terminal indecision." He reinforced this perception by once saying, "When it comes to some of these complex pieces of legislation, I wish that we didn't just have a choice of saying 'yes' or 'no.' I often wish that we had a column marked 'maybe.'" He also received criticism for ineffective questioning of those testifying before committees. As a way of demeaning the inexperienced senator, some sarcastically called him "the best astronaut in the Senate," at a time when he was, of course, the only astronaut in the Senate.[32] Glenn proved to be a traditional liberal on social questions; however, he followed his own drummer on fiscal and foreign policy issues, where he was more often rated as a centrist.[33]

One of the first problems he tackled was the energy crisis of the 1970s. In a decade that had seen gasoline rationing as well as long lines at the pumps, Glenn served on the Senate Energy Committee headed by Henry "Scoop" Jackson, another Democrat. Glenn helped manage a major energy bill on the Senate floor in April 1975.[34] A significantly amended version passed and was signed by President Gerald Ford in December after almost a year of debate between the White House and congressional Democrats. The final bill was intended to put a ceiling on crude oil prices. It contrasted sharply with Ford's own proposed policy, which would have allowed sharp hikes in consumer prices. Ford had hoped that increased prices would generate higher domestic oil production and encourage consumers to practice conservation.[35]

After a little more than a year in the Senate, Glenn became the subject of a *New York Times* analysis. "The first clue that John Glenn had no intention of drifting along on past accomplishments came early in his Senate career when he turned down assignments on the

Aeronautical and Space Sciences and Armed Service Committees," wrote Marjorie Hunter. "He has worked hard, and colleagues give him good marks for his efforts on energy and nuclear matters." Fellow lawmakers rated poor performances in hearings as his biggest flaw as a senator.[36] (Glenn later decided that joining the Foreign Relations Committee instead of the Armed Services Committee had been a mistake.)

The year 1976 opened the door to the possibility of claiming a higher office. Jimmy Carter—the epitome of an anti-Nixon—led the race to the Democratic presidential nomination. This churchgoing populist candidate appealed to voters who were sick of Watergate and disappointed in President Gerald Ford's decision to give Nixon a blanket pardon after his August 1974 resignation. Carter's only elected offices had been as a Georgia state representative and governor.

A staunch Southern Baptist, Carter reached out to small-town people and campaigned vigorously in the many Democratic primaries. During the campaign, Carter visited competitive states early to claim voters before his opponents came onto the scene. In the South, he stressed his regional background, and in the rest of the country, he appealed to religious and rural voters. His lack of Washington ties was a definite plus in his battle for the nomination. Initially considered a dark horse, he moved into the lead after winning the Iowa caucuses and the New Hampshire primary. Running as a man of the people, his informal connections to voters came in sharp contrast to Nixon's imperial presidency.

Carter's campaign identified Glenn as one of ten possible running mates. Carter staffers produced reports on Glenn and other potential vice presidential candidates—analyses of their likely effect on Jewish voters, studies of past campaign finance records, and tallies of their ratings by the liberal Americans for Democratic Action and the AFL-CIO's Committee on Political Education. In Carter campaign interviews, allies described Glenn as "a shy deferential man, who is trying hard to become a good Senator." They classified him as a "true liberal" and "the nicest, best-liked man in the Senate." Others criticized him for indecision on hot issues.[37] In the analysis of Jewish voters, he did not rank among the three "positive" contenders, and instead fell into a five-man group of candidates who were rated as "neutral"—unlikely to attract or repel Jewish voters.[38] His Senate campaign had received three donations that the Carter team considered questionable,[39] and Glenn was in the middle of the pack on both the ADA and the COPE ratings—a mild liberal.[40]

In looking for a running mate, Carter supporters felt that religion was an important issue. They feared anti–Southern Baptist zeal, much as John F. Kennedy had faced anti-Catholic prejudice. They planned to fight religious bias much as Kennedy had—by confronting it openly and by choosing a running mate who could deliver states into the win column. Longtime broadcast executive Robert E. Kintner, who headed first ABC and then NBC between 1949 and 1965 and served as Lyndon Johnson's cabinet secretary, wrote a memo on the contenders for Hamilton Jordan, a key adviser within Carter's campaign. In that note, Kintner advised that Maine senator Edmund Muskie's greatest asset was being a northern Catholic experienced in government. Kintner believed that Glenn had the same type of hero worship that made Dwight D. Eisenhower so popular politically. He did not think Minnesota's senator Walter Mondale would add much to the ticket and advised that if Carter wanted a liberal vice presidential candidate, he would find Idaho senator Frank Church more helpful. Kintner also said Henry Jackson was a decent example of a more conservative Democrat, although he did not have great political appeal.[41]

As the Democratic National Convention began in New York City, Carter aides reportedly cited Glenn as one of the top two choices, and yet, the *Los Angeles Times* reported, Glenn's chances had been diminished after his personal interview with Carter the previous week. Apparently, the presumptive presidential nominee was not impressed with the former astronaut.[42] On the convention's first night, Glenn was scheduled to deliver a keynote address. Years later, newsman Walter Cronkite summed up Glenn's performance well: "That was the most impressive thing about that convention to me. It was how John fell on his face. It was sad. Everybody thought he was presidential timber, and he made just one of the world's worst speeches. It was just ghastly."[43]

(Interestingly, a strong keynote at the 2004 convention transformed Barack Obama from a relatively obscure community activist and state legislator to a nationally known figure who would capture the presidency just four years later.)

Delivered in what one reporter called "an almost conversational tone," Glenn's speech created little excitement. His overall message was a return to the "ideals, beliefs, and confidence" that once defined the United States. While he spoke, Madison Square Garden was filled with noise as Democrats were "milling about, bumping into television cameras, paying little heed to the early speakers," the *New York Times* reported.[44] Here's an excerpt from Glenn's address: "No haphazard drift to the future this, but pride and accomplishment together, a pride not passé, a pride not old-fashioned, but pride in a new patriotism, a patriotism based on Participation, a patriotism based on Accomplishment."[45]

Right-wing columnist Patrick Buchanan wrote, "Though his audience was discourteous and uncivil, Ohio Sen. John Glenn's keynote address to the Democratic National Convention was singularly uninspiring. So, contrasted with oratorical flourishes of convention favorite Rep. Barbara Jordan of Texas, America's astronaut-hero saw his vice-presidential hopes fizzle there on the launching pad."[46]

After the speech, Glenn was unhappy to think that his speech had destroyed his chance at the vice presidential nomination, but he was more troubled by a rumor he had heard. "The only other thing we ever heard and it was never repeated, was that Rosalynn Carter, Mrs. Carter, thought that with Annie's speech problems, that that would not be too appropriate for a First Lady. In fact, we asked . . . [Carter's press secretary] Jody Powell . . . and he wouldn't answer me on that." This suggestion saddened the Glenns.[47] Carter, with Mondale as his running mate, went on to defeat Gerald Ford, who had risen to the presidency after becoming the United States' first appointed vice president.[48]

In the Ninety-Fifth Congress, which served from 1977 to 1979, Glenn introduced twenty-two bills and cosponsored fifty-seven more. They addressed some substantive issues, but many fell into the category of minor suggestions, such as paying tribute to Alex Haley for the success of his book, *Roots*, and a request for the president to proclaim October 7, 1978, as "National Guard Day."[49]

Christine McCreary joined Glenn's office staff in 1977 and stayed on until his retirement in 1998. In a 1998 oral history, she said, "Senator Glenn is so easy-going and so for real that it's just like you could be talking to a friend or a brother.... If he sees me in the hall, he hugs me. He's just that type of person and Mrs. Glenn, too." She called Glenn's Senate office operation "smooth-running" and added, "Everything is togetherness as far as he's concerned in the office." Nevertheless, she made it clear that he did not accept foolish behavior. "He doesn't play around.... You do some silly, stupid things, you're gone."[50]

Another longtime employee was Kathy Connolly, who was with him almost from his first election to his retirement and said that in all that time, "I never even ever remember him raising his voice." Working in Glenn's office was exciting for her. "It was my first job out of college. I worked in the mailroom, and I ended up for many years working as a legislative assistant for him." She assisted Glenn in addressing issues tied to agriculture and social concerns such as education. Connolly recalls Glenn's office as a workplace where women received opportunities and where children played when their mothers had to work at night. She remembers one night when she briefly lost track of her three-year-old son and found him sitting with Glenn in his office, where they were sharing a cache of popcorn.[51] At least once during his Senate career, Glenn rolled up his sleeves and used his jumper cables to start a secretary's car.[52]

By far, the issue with which Glenn became most closely connected in the 1970s was nuclear disarmament. He sponsored the Nuclear Non-Proliferation Act of 1978, which won Senate approval. "Our objective at that time was to try and prevent the spread of nuclear weaponry to more and more people around the world," he said. The five openly nuclear nations at that time were the United States, the United Kingdom, France, the Soviet Union, and China. While the treaty represented a significant achievement in minimizing the dangers of nuclear war, Glenn later admitted that the measure was not entirely successful. Some other nations, such as India, subsequently showed off their nuclear weapons programs.[53] (Pakistan also has developed its own nuclear weapons program.) Nonetheless, the General Accounting Office concluded in 1981 that the act represented "an important step by the United States to establish a framework of controls and incentives that, if adopted internationally, could reduce the threat of weapons proliferation and promote the peaceful uses of nuclear energy." The GAO also recommended periodic alterations to reflect changing political, economic, and technological realities; to draw wider support from other nations; and to make implementation more effective.[54] Glenn said in 2009 that, over the years, diplomacy managed to delay and halt nuclear weapons programs in Argentina, Brazil, Taiwan, and South Africa. Israel never admitted to having a nuclear program, but Glenn said, "everyone knows they do."[55]

Much more controversial was his stand on President Carter's Strategic Arms Limitation Talks, called SALT II, an agreement with the Soviet Union in 1979. (The Nixon administration had negotiated SALT I.) Much to President Carter's dismay, Glenn opposed the new arms control treaty even before it had been fully negotiated. Glenn's objection to the treaty was the United States' current inability to confirm Soviet compliance. Previously, the United States had used Iran as a base for spying on Soviet weapons development. That arrangement died when a revolution toppled the shah of Iran in 1979 and eliminated Iran as a possible site for US observation. Glenn believed that the treaty was intrinsically flawed and stubbornly stood against it.[56] He wrote that his position on this issue "led to a phone call from the president that developed into a harsh exchange. No president before or since has ever talked to me that way, and I've never spoken that way to any other president. Carter said I could kill arms limitation while the treaty was still being negotiated."[57]

After final talks between the United States and the Soviet Union, both Carter and Soviet general secretary Leonid Brezhnev signed the agreement in June 1979, and Carter sent it directly to the Senate for approval. The final treaty would have limited the nuclear arsenals of the United States and the Soviet Union to 2,250 delivery vehicles and placed other restrictions on nuclear weapons.[58] Carter promised that satellite reconnaissance and new observation locations would soon give the United States solid information about Soviet weapons advances; still, Glenn refused to support the treaty in the Senate Foreign Relations Committee.[59] One critic said it was not just that Glenn couldn't see the forest for the trees: "He looks at the bark on the trees." Some felt that his penchant for focusing on detail made him a "stubborn and bull-headed"[60] lawmaker unwilling to contemplate the possibility that even an imperfectly enforceable treaty could promote disarmament.

Despite Glenn's opposition, the Foreign Relations Committee approved the treaty by a vote of 9 to 6 in November 1979.[61] A month later, the Senate Armed Services Committee issued a report condemning the treaty, and nineteen senators urged Carter to delay the treaty until military concerns could be thoroughly reviewed.[62] Nevertheless, some of the military's strongest backers in the Senate supported the treaty. The year ended without approval of the treaty, and in January 1980, Carter himself asked the Senate to delay consideration following a Soviet invasion of Afghanistan. Later in 1980, Carter said that he would follow the unapproved treaty guidelines if the Soviet Union would make the same promise. In 1986, Carter's successor, President Ronald Reagan, declared that the Soviet Union had not kept its commitment and abandoned the policy of following the principles laid out in the still-unratified treaty.[63]

In addition to disagreeing with Carter on the SALT II treaty, Glenn opposed Carter's proposed withdrawal of six thousand troops from South Korea and his cancellation of the B-1 bomber's production. Glenn was chairman of the Asian Subcommittee of the Senate Foreign Relations Committee, so the White House gave serious consideration to his concerns about the Korean pullback[64] and his request for a one-year delay. Subsequently, no significant pullout occurred.[65] The B-1 bomber later was brought back into production by Reagan and continues to fly three decades later.[66]

In 1976 and again in 1979, Glenn visited the People's Republic of China. He made the first trip with Senate majority leader Mike Mansfield just a few weeks after Mao Zedong died. Glenn was hesitant to travel and miss roll-call votes; however, he decided to take the trip. The lawmakers' plane was allowed to land in Beijing with directions that it must take off and leave the country immediately after the passengers deplaned.

In China, Glenn was appalled to learn that the huge nation was not yet fully electrified. At various stops, Glenn asked who was planning the nation's electrification, and the answer was always the same: "The thoughts of Mao and the will of the people." The lawmakers also got a chance to watch acupuncture in action and to ask questions about the technique.[67] On his second trip in 1979, with a delegation that included Democratic senators Robert Byrd and Gary Hart as well as Republican representative John Anderson, Vice Premier Deng Xiaoping demonstrated that Glenn's reputation as a spaceman was not limited to the United States. He asked, "Do I understand that Senator Glenn was on the moon?" After Glenn correctly identified his space mission, Deng said, "In China, you would be called a person of the other world. This means that a person, a mortal person, has come down from the celestial world." Glenn responded glibly, "My constituents call me many things, but that's not one of them."[68]

By 1980 it had become clear that while being an "outsider" may have made Carter successful as a candidate in 1976, it had lessened his effectiveness as president. When both Glenn and Carter faced reelection, Glenn won big and Carter lost big. Glenn trounced his Republican opponent by more than 1.6 million votes. Reagan claimed the presidency by winning more votes than the combined totals captured by Carter and third-party candidate John Anderson. Carter carried only five states plus the District of Columbia and won only forty-five electoral votes.[69] He became the first elected president since Herbert Hoover to lose a reelection bid.[70]

Glenn's large victory in Ohio was especially notable because Reagan had carried the state by a four-hundred-thousand-vote margin. This impressive victory set the stage for Glenn to assume a higher national profile and to examine the possibility of seeking the presidency.[71]

11

THE EBB AND FLOW OF POLITICAL LIFE

After winning a landslide victory in his 1980 Senate bid for reelection, Glenn found himself blessed with a $100,000 campaign surplus and growing interest in the possibility of a run for the presidency. The senator had known several presidents personally, and through his close friendship with Robert Kennedy, he understood that presidential candidates and presidents were not distant, august figures; they were human beings like him, and knowing that probably made it easier to imagine seeking the presidency himself. As a young man, he had been bold enough to consider the possibility of being commandant of the Marine Corps; this goal, like that one, was a reach—but not an impossible one.

After Reagan's 1980 victory, Democratic leaders believed that Senator Edward Kennedy would be the front-runner for the Democratic Party's nomination in 1984. Since Robert Kennedy's death in 1968, the youngest Kennedy brother had been a powerful force within the party, although he lost a challenge to the incumbent Jimmy Carter for the 1980 Democratic nomination. While his name and face generated favorable responses, these feelings were accompanied by suspicions surrounding Mary Jo Kopechne's 1969 death inside a car he was driving on Massachusetts's Chappaquiddick Island. After careful consideration, Glenn decided to begin exploring the possibility of his own candidacy. He and his aides made their first visit to New Hampshire in 1981, and in June 1982 they participated in the Democratic midterm conference in Philadelphia. All of the eventual contenders except Jesse Jackson and George McGovern participated in the midterm gathering, where potential delegates could be wooed.[1]

Looking to the future, Glenn never failed to remember his past and always remained a strong advocate of NASA. Just a few weeks before the April 12, 1981, launch of the first space shuttle flight, Glenn spoke to the National Space Club and cited the many ways that space exploration had benefitted Americans in their daily lives. He credited satellite technology with saving thousands of lives through storm tracking, making international phone calls less expensive, and improving the quality of international TV reception. Creation of the integrated circuit board for space travel, he pointed out, had made it possible to engineer digital watches, handheld calculators, and newly emerging but not yet ubiquitous desktop computers. Moreover, computerized analyses made buildings, ships, cars, and planes more structurally sound.[2] A year later, he urged Senator Jake Garn, chairman of the subcommittee that oversaw NASA funding, to increase financial support for the agency, regardless of Reagan's drive to cut taxes and slash federal budgets.[3] Throughout his Senate career, "Glenn was vociferous in trying to say you just couldn't allow across-the-board cuts, that you needed to make cuts based on some rational decision-making," said legislative director Ron Grimes.[4]

As he worked in the Senate, talk about Glenn's presidential qualities buzzed around him. A 1982 *People* magazine article described his appeal as bipartisan. "Glenn has the star quality, but the jury is out on his ability to organize," said Representative Tony Coelho of California, chairman of the Democratic Congressional Campaign Committee. Even disgraced GOP president Richard Nixon sang Glenn's praises, saying he "comes through as a very responsible man with some new ideas. . . . The man with new ideas is the only one to give Reagan a race."[5]

Glenn's chances soared in December 1982, when Kennedy announced that he would not run, citing an "overriding obligation" to his three children, then twenty-two, twenty-one, and fifteen. The *New York Times* reported that Glenn's stock "took a leap" with Kennedy's withdrawal, and colleagues in a Senate hearing "teasingly addressed him as President Glenn."[6] Suddenly, something that had seemed remote became much closer to reality. Glenn began working with a coach to improve his often harshly critiqued public speaking. He had wowed the Congress in 1962 after his orbital flight, but the 1976 Democratic Convention keynote debacle was a fresher memory and more typical of his style when not wearing the aura of a hero.

A December 1982 Gallup Poll showed Glenn taking a strong lead over Reagan—54 to 39 percent; the other Democratic front-runner, former vice president Walter Mondale, led the president 52–40 at a time when Reagan's approval rating had plummeted to its lowest level ever, 41 percent.[7] The polls encouraged Glenn to become increasingly focused on the presidential sweepstakes.

Both Glenn and Mondale faced the challenge of raising a lot of money during 1983—not a task Glenn enjoyed. "I'd rather wrestle a gorilla right here on the floor rather than ask anybody for a couple of bucks," he later said.[8] Mondale, on the other hand, had strong ties to labor unions and other special interest groups that could fill his campaign coffers. Meanwhile, outside sources began to weigh in. The League of Conservation Voters gave Glenn only "a mediocre environmental record" in the Senate in March 1983.[9] In the next two months Senator Paul Tsongas of Massachusetts lent his support to Glenn, and Mondale claimed the backing of a long list of public officials.[10]

In April, Glenn officially announced his candidacy in New Concord at John Glenn High School. Some questioned the decision to make his announcement so far from the media spotlight, but Glenn later argued, "I thought about the things that we had talked about here, and I thought everything is not based in the cities. I thought it was important that we included all of America and that people know that starting out."[11] Glenn traveled over the next six days from Washington to Los Angeles, San Diego, Houston, Little Rock, and Dubuque.[12]

Weeks later, a *Washington Post* report found that Glenn's annual earnings were the highest among Democratic candidates. His income was between $1.6 million and $2.1 million, and his assets were valued as high as $7.4 million, also exceeding the asset figures for the other active candidates—Mondale, Senator Alan Cranston of California, former Florida governor Reuben Askew, Senator Gary Hart of Colorado, and Senator Ernest Hollings of South Carolina.[13] (Jesse Jackson did not enter the race until November 1983.) In a May *Los Angeles Times* poll, Glenn polled even with Mondale for the first time among Democrats.[14] The following month, the *New York Times* speculated about Glenn's potential to reproduce GOP president Dwight D. Eisenhower's heroic, avuncular appeal. Humorist Mark Russell joked that Glenn "even looks like a Republican."[15]

Ike's shadow would follow Glenn throughout his campaign; the bald head, the graying fringe, and the cloak of righteousness all made the comparison obvious, but changing times made the 1984 campaign more contentious than the races Eisenhower faced in the 1950s. A July *New York Times* / CBS News Poll of Democrats showed Glenn with a 53 percent favorable rating and only an 11 percent unfavorable rating. Mondale received a 51 percent favorable response, while 19 percent viewed him unfavorably. In third place was Jesse Jackson, who had not yet announced his candidacy, with a 23 percent favorable rating and a 26 percent unfavorable score.[16]

A study prepared for the Glenn campaign dated June 1983 found that Americans interviewed in focus groups had general ideas of Mondale as a former vice president and Glenn as an astronaut but few knew the political positions taken by either man. In general, voters viewed Mondale as a "politician," while they saw Glenn as a "moral leader." Each man needed to clarify his policy objectives.[17] Several months later, when Glenn's position papers emerged, they did not immediately define his positions. Instead, they began with uncontroversial, nonspecific quotes from the candidate. For instance:

Economy: "What I offer is not the false hope of supply-side economics—or the empty promise of reborn big spending. Instead, I offer a program that is grounded in reality—and based on common sense."[18]

Foreign trade: "Our problems in foreign trade lie not with our people but with the absence of leadership. We're the same people who tamed a continent, crossed frontiers and built the greatest, strongest nation on earth. We see a problem, think up an idea, test it, adjust it, and finally succeed with it. And with the right leadership, we can still outwork, outinvent, outproduce, and outcompete any nation on Earth."[19]

Industrial policy: "America has faced and overcome difficult challenges before, and we can do so again. Many of today's problems are the result of tremendous global change. But change is the law of life—and it brings opportunity as well as danger."[20]

Agriculture: "Farming is still this Nation's greatest success story, and the efficiency and productivity of our farmers remain the envy of all the world."[21]

Energy: "When it comes to meeting the energy challenges of the future, it's not enough to promise to get government either off your back or on your side. Ideology won't make us energy independent, and rhetoric can't fuel our economy."[22]

In short, the message was: if it's broken, we can fix it. Instead of introducing specific solutions, these speech excerpts offered no clue about how he would tackle any of these issues. None of the quotes takes a firm stand—and the policies explained beneath them are filled with vague promises of developing policies to correct current conditions rather than specific steps to solve problems. Taking an unequivocal stand requires greater conviction than generalizations that convey no information. Some observers wondered whether Glenn even had a vision to share. "He really doesn't seem to have a grand design for the America he'd like to see," one lobbyist told the *Washington Post*, and a former Glenn aide said, "He's the original clean slate."[23]

Dale Butland viewed it differently. "It is very difficult to generate passion in a primary if you are from the sensible center," he said. Centrist policies seldom offer the drama of more extreme proposals. "Primaries are always controlled by party activists, who tend to be on the left if you're a Democrat and on the right if you're a Republican."[24]

For the candidates, 1983 represented a long buildup to an election year. It was filled with diversionary events that distracted Glenn from fundraising. "Glenn's people had been bickering over where to go and what to do. Mondale's had accurately seen 1983 as a prolonged first primary—a shadow war whose objective, for them, was to heighten the perception that Mondale was an irresistible force," wrote Peter Goldman and Tony Fuller in *The Quest for the Presidency 1984*.[25] There were straw polls at various state Democratic events, and candidates had to choose between fundraising events and these gatherings, which guaranteed no delegates but offered candidates an opportunity to garner attention and enhance their apparent standing. Sometimes individual groups within the states' parties scheduled events to meet with the candidates while they were in town for the straw poll. Glenn participated in many of these contests, although he announced in late July that he would not invest a great deal of time or effort into the straw polls.[26]

On October 1, 1983, Mondale captured victory in the Maine straw poll with a majority of the vote in a contest that included three other candidates, and within a few days, he received a boost from two major endorsements—one from the National Education Association and another from the AFL-CIO.

There were rumors about trouble in the Glenn campaign almost from the beginning. "We rolled that baby out of the garage, and the wheels fell off," said Butland.[27] Glenn's critics sometimes reported that he was MIA at key moments. Campaign watchers noticed that he was the only candidate to miss the Massachusetts reception before that state's straw vote. He had given higher priority to a commitment in Ohio. Furthermore, he failed to attend a breakfast the next day hosted by black and Hispanic delegates. One Glenn campaign spokesman justified his absence, saying that he wanted to spend more time talking to individual delegates, while another adviser described it as "a gaffe." Columnists Rowland Evans and Robert Novak noted that part of Glenn's problem was the absence of valuable endorsements from his home state, and that prompted Glenn to pick up the phone and call the three key people mentioned. Representative Edward F. Feighan was one of the three. He was a freshman representative from Ohio. When he won his congressional election, Mondale had made a congratulatory call, but Glenn had not.[28] To an experienced politician like Mondale, touching base with a newly elected Democrat was an obvious friend-raising tactic; to Glenn, it apparently was not. Even after more than a decade in the Senate, he had not mastered the political tools of the trade. In truth, being called a great politician was not an honor he ever craved. Acting as a public servant was one thing; being a politician was something else.

Over the course of 1983, Glenn found that Americans' tendency to see him first and foremost as an astronaut was a mixed blessing. It positioned him as a single-combat hero but offered no proof that he could lead others. It was difficult to minimize that part of his life, especially late in the year, with actor Ed Harris portraying him in *The Right Stuff*, a major motion picture based on Tom Wolfe's book that was debuting on movie screens. The film made Glenn's astronaut role more than an indelible memory; it guaranteed that Glenn's astronaut past would maintain a place in American memory and popular culture.

Nationwide polls in 1983 showed Mondale leading Glenn 42 percent to 18 percent in October, but a November *Los Angeles Times* poll found Glenn two percentage points ahead of Mondale. Gallup showed Glenn with a seventeen-point advantage over Reagan. These November polls sparked a fundraising push in which the Glenn campaign sought additional donations from forty-eight hundred previous contributors.[29] However, winning over new contributors was an uphill battle. *Newsweek* reported that Mondale conjured nostalgia for old-time liberals like Hubert Humphrey, but "Glenn appeals to the Democrats' it-takes-a-hero-to-beat-a-movie-star pragmatism."[30]

The *New York Times'* James Reston, who had known the former astronaut for decades, wrote in November 1983 that no matter what the polls said, Glenn could not compete with Mondale. "He's an intelligent, dead-honest character, a middle-of-the-roader, a bit of a 'square,' a Presbyterian in a secular age."[31] By the conclusion of 1983, Glenn had raised $5.7 million, putting him second only to Mondale, who had collected almost twice as much. In late 1983 a *Boston Globe* poll of New Hampshire primary voters showed Mondale winning 46 percent; Glenn, 16 percent; and Gary Hart, 8 percent.[32] Virginia governor Charles S. Robb joined Glenn's team in December amid rumors of a possible vice presidential nomination for Robb.[33]

As the contest became more competitive, campaign rhetoric changed. Until that point, Glenn had refused to make personal attacks on his opponents or to criticize their stands on the issues. Avoiding backstabbing verbal assaults on his opponents was commendable, but the issues were important to voters. When questioned in one of several New York forums hosted by Governor Mario Cuomo, Glenn declined to outline his disagreements with Mondale's positions. The former vice president, whose campaign was becoming concerned about the threat Glenn posed, tackled the question very differently in another of these gatherings. Mondale pointed out their differences very specifically and succinctly, while being careful to make his stands sound correct. "I strongly supported SALT II and he opposed it. I'm opposed to the B-1 bomber, and he favors it. I wrote a letter to the Senate urging that they turn down poison nerve gas, and he voted for it. I believe in real increases in the defense budget, but at a reasonable, manageable pace. The figures he proposes are substantially higher than those that I think are supportable."[34] Glenn was frustrated by Mondale's decision to make the contest personal and to cast his stands in a negative light without providing what he considered to be proper context. To counter Mondale, he pulled a charge out of his back pocket that he had been saving. He saw Mondale's dependence on interest groups as his Achilles' heel, so he asked, "Will we offer a party that can't say 'no' to anyone with a letterhead and a mailing list?"[35]

Ironically, Glenn emerged from the New York forums wanting to battle Cuomo as well as Mondale. Though he was not a candidate, the governor had entered the political in-fighting by calling Glenn a "celluloid candidate" as a result of *The Right Stuff*'s release. Glenn countered Cuomo, saying, "I wasn't doing *Hellcats of the Navy* [a Reagan film] when I went through 149 missions. That wasn't celluloid. That was the real thing. And when I was on top of that booster down there getting ready to go, it wasn't *Star Trek* or *Star Wars*."[36] (Glenn believed *The Right Stuff*, which won four Oscars, plus a Best Picture nomination, offered entertainment but had no connection to his political success or failure.[37] He later told a *New York Times* reporter that he did not like the "square goodie-goodie" portrayal of himself.)[38]

From the start, Glenn's campaign had been disorganized, filled with internal disputes and too many would-be leaders. "Our presidential campaign in '84, like many others, was basically run by a bunch of hired guns—professional political actors who were not necessarily Glenn loyalists," Butland said. In fact, these outsiders did not demonstrate a clear understanding of the senator's political stands. Instead of capitalizing on the senator's 100 percent civil rights voting record, they placed Glenn in front of a Confederate war memorial in Georgia for a news conference, much to the dismay of both Glenn and Butland.[39] The senator realized that his campaign was in chaos, but "every time I wanted to come off the road and said that I wanted to go back into Washington for a week and sit down and reorganize this whole thing, by then we had so many events planned. . . . And so I never did come off the road, really, and try and correct the set-up that we had."[40] In other words, the mess made fixing the mess impossible. On the campaign trail, Glenn connected with voters best in one-on-one chats when he stopped to sign autographs or to ask a voter his or her name. As a speaker, he still was not dynamic.

The closely packed primaries and caucuses of 1984 offered little opportunity to bounce back from a loss before facing a vote in another state, and Glenn's strategy raised questions among journalists and politicians. Glenn adopted an unconventional game plan without studying the tactics that had brought success to previous contenders.

Rather than adjusting his pitch specifically to win over voters in Iowa's caucuses and in New Hampshire's primary—the crucial first two stops on the road to the nomination—Glenn concentrated on framing a national campaign with Super Tuesday primaries in Alabama, Georgia, and Florida as his most important targets. In Iowa, instead of being an ever-present and visible candidate, he sank money into TV ads, which might have worked in other states but wasn't a viable strategy in Iowa, where voters were accustomed to personal contact.[41] He skipped an all-candidate debate on farm issues as well. At the Jefferson-Jackson dinner in Iowa before the caucuses, Mondale claimed that it was Glenn, not he, who represented special interests. He contended that Glenn's support for Reagan's budget and tax cut in 1981 had benefited the wealthy. (In reality, only seven Democrats voted against Reagan's tax cut in its final form.)[42] As the last speaker, Mondale attacked Glenn on a variety of fronts, and Glenn could not respond.

As the campaign neared the first real contests, potential fundraising opportunities, not early voting priorities, guided Glenn's speaking schedule.[43] Reporters in Iowa noticed that he was missing and that his headquarters seemed noticeably empty compared to those of his rivals. "We did not concentrate enough on Iowa and New Hampshire," Glenn recalled years later.[44] Rather than following his competitors' example and contacting previous caucus participants, Glenn's campaign struggled to build a new constituency of previous nonparticipants as opposed to letting liberal party activists lead the charge; broadening the electorate was seen as a requirement for a decent showing.[45] This strategy did not remake the electorate. The *Washington Times* reported, "The Ohio Democrat's views may be in tune with the views of the majority of Democrats nationwide, but they seem to turn off too many of the activist Democrats who decide who will be the party's standard bearer."[46]

At the Florida Institute of Technology, Glenn tried to shoot holes in Mondale's bill of indictment against him, but he went a step too far in the minds of many Democrats. Instead of strictly attacking Mondale, he expanded his criticism to include Jimmy Carter. Targeting the only living Democratic former president was considered bad form, even when that president was not terribly popular. Glenn said that in Reagan's 1980 victory over Carter, "the people had voted their fears." He went on to identify Mondale as "part of the administration that gave us 21 percent interest rates and 17 percent inflation rates."[47] Each time that Glenn made a harsh attack on Mondale, he did so at the expense of his heroic image. "What happened to John Glenn's campaign for president is that it took a candidate with the image of Dwight D. Eisenhower and turned him into Rocky Graziano," reported the *Washington Post*.[48]

When it came time to count the votes in the Iowa caucuses, Glenn got only 3.5 percent of the vote and finished a dismal sixth—after four candidates and "uncommitted."[49] Gary Hart, who finished second, stole Glenn's spot as Mondale's strongest competition. In-fighting had crippled the Glenn campaign, and leadership went from one man to another as the candidacy drifted toward oblivion. Iowa had only 58 of the more than 3,923 delegates to the Democratic National Convention,[50] and yet, it had already played a major role in reordering the cast of contenders.

After the Iowa caucuses, only eight days remained before the New Hampshire primary, where Hart stunned Mondale, winning with 37.3 percent of the vote. Mondale landed in second place with 27.9 percent, and Glenn finished a distant third with only 12 percent.[51] In an analysis of media campaign coverage, Keith Blume argued that journalists "at first pumped" Glenn, then "abandoned [him] for not having the right stuff for TV politicking."[52]

Next up were the Super Tuesday primaries on March 13, where Hart won seven events and Mondale took two. Glenn withdrew from the race two days later. Kathy Connolly, who left her Capitol Hill job to work for the campaign in New Hampshire and Alabama, said he left the race "with class and dignity and a lot of congratulations to what is going to be ahead."[53] Mondale's strategy of attacking and belittling Glenn worked to minimize the former astronaut's chances but did not guarantee the former vice president's hoped-for sense of invincibility.

Glenn conceded twenty-five years later:

> When I look back on the national campaign, we made a lot of
> mistakes. We were not able to raise the amount of money that
> you needed for something like this. . . . I wanted to run this as
> a whole national campaign. I still think to this day, that to let
> Iowa and New Hampshire almost dictate what happens in the
> campaign and be the big bellwether of everything that's going to
> happen all across the country, I think is not the way we should
> be doing this. . . . If someone in Iowa and in New Hampshire . . .
> had not met you two or three times personally, they think you're
> ignoring the state or something. I thought this was ridiculous
> then, and I still do.[54]

Glenn had no regrets about the decision to run, but he said, "When I
think about it myself sometimes, I think maybe it was Don Quixote,
jousting against windmills or something."[55] And he knew what the
most basic problem was: "I may not like the traditional steps of Iowa
and New Hampshire, but that was it."[56]

Bad news did not end with the close of the campaign. The *San
Francisco Examiner* conducted an analysis of reckless spending within
the Glenn campaign. Based largely on interviews with Glenn staffers,
the investigation found that while the campaign was running low on
supplies to mail out campaign literature to voters in the days leading up to the New Hampshire primary, organizers invested $500 in a
cake shaped like the United States and spent $30,000 on hotel rooms.
High-ranking campaign officials apparently traveled in luxury, while
office staffers sometimes scrounged for staples and paper clips. Steve
Avakian, who managed Glenn's 1980 Senate race, reported that the
campaign experienced spending abuses that may have been "unprecedented in American politics." Glenn advance men carried handfuls
of bank drafts redeemable for $250, $500, and $1,000. In addition,
the investigation identified situations in which campaign money was
used for personal expenses. Glenn staffers indicated that no one kept
a close accounting of individual expenses by collecting receipts or logs

of expenses. A Mondale staffer said, "We were aware of some surprisingly robust salary offers. They were moving very fast, and money was seemingly no object."[57]

"The financial chicanery was just shy of criminal," said Butland, who credits the hired guns with much of this problem, but conceded that one of Glenn's weaknesses "was that he was not a particularly good manager. . . . He just depended on other people to do the right thing." In the presidential campaign, he apparently relied on the wrong people. Butland recalled "outrageous stories like, for example, the Iowa campaign staff flying ribs in from Kansas City so they could have a ribs fest." And after the New Hampshire primary, he said, "We found thousands of dollars' worth of postage stamps that were left in desk drawers" in the campaign office.[58]

Glenn ended his run for the presidency with a massive debt of $3,632,194.23.[59] "There is nothing quite as unattractive as a just-defeated candidate as far as financial contributions go, so that was very difficult raising money," he later said.[60] Nonetheless, he staged fundraisers in hopes of filling the gap. By 1991, Glenn still owed more than $2.75 million, as interest rates kept the dollar figure high even after money had been paid to lessen the debt.[61] The debt troubled Glenn. "I had already at that time put in all I could put in legally. A candidate is limited." He petitioned the Federal Election Commission and was given permission to contribute more of his personal wealth to pay off small vendors. Then, after many years, the commission allowed the four banks owed money to write off the unpaid remainder, Glenn said.[62]

Following her husband's withdrawal from the presidential race, Annie, whose confidence in public speaking had improved greatly, spoke to the Platform Committee of the Democratic National Committee in May 1984. Her topic was one she knew well: the plight of Americans with disabilities. After many years of struggling with stuttering, she told the committee, "The most important part of any community is people—and I know that far too often, disabled Americans are still denied the opportunities they need to become full-fledged members of our communities. And when that happens, it is not just the handicapped who suffer. In a larger sense, we all do—because we lose the contributions these people could otherwise make to our cities, our states, and our society."[63]

Glenn quickly began returning his attention to Ohio, with an eye on his planned 1986 reelection campaign. Butland recalled that Glenn often told self-deprecating jokes about his dolorous presidential campaign:

> You know, this campaign didn't turn out the way I hoped it would, but it was all Annie's fault. For years, Annie told me she wanted me to run for president in the worst way possible. And that's exactly what I did. . . .
>
> But I had the full support of my family. . . . Even my son Dave took time off from his medical practice to help out. Dave is an anesthesiologist. He puts people to sleep. And you know what they say: like father, like son.[64]

On the still-contentious campaign trail, Mondale, Hart, and Jackson split the remaining primaries. Hart's campaign entered the convention believing that a shift of seventy-five votes from Mondale could give Hart the nomination, but Mondale hung on to win and partnered with the first female vice presidential candidate nominated by one of the major parties. In November, Mondale and Representative Geraldine Ferraro of New York lost the election badly, claiming only thirteen electoral votes, while Reagan and George H. W. Bush won 525.[65] Some Mondale backers claimed that the fierce attacks of Glenn and Hart during the primary season had weakened the former vice president's candidacy.

His presidential dreams behind him, Glenn returned his attention to the Senate in 1985 and decided to make a change. He chose to leave the Foreign Relations Committee and join the Armed Services Committee, where he expected to be more useful. "Frankly, the Foreign Relations Committee wasn't doing much of any consequence that I could figure out," he said.[66] Despite ten years in the Senate, he began his service on the Armed Services Committee without seniority.

NASA had a rough year in 1986, and Glenn still shared in the agency's failures as well as its successes. On January 28, the shuttle *Challenger* broke apart in a fiery blast seconds after liftoff. All seven astronauts, including teacher Christa McAuliffe, were killed by depressurization or by impact with the ocean when the crew cabin crashed at more than 200 mph. These were the first American astronauts to be killed during a NASA mission. This first fatal space-mission accident could have happened in a one-person, two-person, or three-person flight, and it could have occurred far out of Americans' view. Instead, seven human beings lost their lives before reaching orbit, and the accident occurred within sight of many thousands of spectators—as well as anyone owning a TV set. (The three major networks were not showing the liftoff live. Only six-year-old CNN, the first twenty-four-hour news network, carried a live telecast, but video of the disaster aired repeatedly.)

Glenn went to the Kennedy Space Center to talk with family members and with NASA employees just hours after the catastrophe. He said, "We have a tragedy that goes along with our triumphs."[67] Glenn took a spot on the board of the Space Shuttle Children's Trust Fund, which had raised almost $1 million less than two months after the accident.[68] The fund, which was expanded to aid children of any astronauts killed during a mission, held more than $2.5 million thirty years later.[69] On the day of the accident, Glenn issued a statement, saying, "In light of National Aeronautics and Space Administration's (NASA) safety record, I have no doubt in the world that this mission suffered something entirely unforeseen."[70]

In this case, Glenn was wrong; the possibility of a catastrophic accident had been predicted by some experts. After a careful investigation, a report concluded that the spacecraft had taken off when the temperature was so low that it adversely affected O-ring joints in the solid rocket booster. The probe concluded that failures in NASA's culture had contributed to the explosion. There were people at NASA who had been warned about the O-rings, but because the temperature warning sprang from a theory rather than hard facts, middle managers did not feel they had sufficient evidence to cancel the launch.

After reelection to a third term in 1986, Glenn became chairman of the Senate's Governmental Affairs Committee in January 1987 as the One Hundredth Congress came into session. Taking over as chairman expanded Glenn's Washington office staff from about thirty-five to approximately seventy. (He had about fifteen staffers in Ohio.)[71] He liked the challenge, although his committee's work struck other senators as uninteresting. "He's a technocrat," one senator told the *Washington Post* dismissively. "He really doesn't like politics."[72] Ron Grimes thought this committee meshed well with Glenn's engineering background: "Much of the Government Affairs Committee did work on the nuts and bolts of where the boxes fit in the organizational scheme. And I always thought that Senator Glenn was an engineer in his thinking and that the way the government boxes lined up made sense to him."[73]

A report of the committee's activities during the first two years of Glenn's eight as chair indicated that the panel's highest priority had been nuclear issues—intensive study of weapons facilities, power plants, and nonproliferation objectives. As a result of the committee's investigations, Glenn proposed an outside safety oversight board for weapons plants; an inspector general for the Nuclear Regulatory Commission, which oversees nuclear power plants; and a legal requirement that the State Department share all information available about nuclear proliferation. Reagan signed all three proposals into law.[74]

In 1988 the possibility of a higher office arose again as the Democrats' presumptive nominee, Massachusetts governor Michael Dukakis, began looking for a running mate. Once more, Glenn's heroic image attracted attention. "It's not just Glenn's resumé—Korean War ace, record-setting jet pilot, the first American to orbit the Earth—though it certainly doesn't hurt," wrote Jon Margolis in the *Chicago Tribune*. "No, it is also John Glenn's face. He has what may be the world's greatest jawline, enough in itself to convince people that he is strong, but not swaggering."[75]

He had backers on Capitol Hill, such as Representative Douglas Applegate of Ohio.[76] One memo warned Glenn that Bush's service in the vice presidency had weakened the job, which had experienced growing power under Vice Presidents Nelson Rockefeller and Mondale.[77] Glenn and Senator Lloyd Bentsen of Texas were finalists. Dukakis apparently chose Bentsen to fortify his showing in the South. A poll of likely voters by the Mason-Dixon Research Group showed Republican nominee, Vice President Bush, trouncing Dukakis in southern states—58 percent to 35 percent. The survey indicated that the possible addition of Glenn to the ticket would have little effect on the southern vote, while putting Bentsen's name on the ballot would cut Bush's margin to five percentage points.[78] Once again a bridesmaid, Glenn focused on his work in the Senate as Dukakis and Bentsen met defeat at the hands of Bush and Senator Dan Quayle of Indiana.

For John Glenn, 1989–1992 represented a ghastly, unexpected period. For the first and only time in his life, he was accused of ethical violations and was forced to defend his reputation. The unraveling of a significant part of the savings and loan industry during Bush's presidency became the catalyst for Glenn's toughest battle to maintain his standing as an ethical legislator. A law signed by Jimmy Carter in 1980 gave S&Ls many of the same powers of banks but without the tight regulation imposed on banks. As a result, some S&L bankers abandoned low-risk home lending to pursue more speculative possibilities. The result was that almost 750 S&Ls failed. Resolution of the crisis cost American taxpayers billions of dollars under a bailout program during the late 1980s and early 1990s. One of the most notorious failures occurred at California's Lincoln Savings and Loan, led by Charles Keating.

Glenn and four other senators—Democrats Alan Cranston of California, Dennis DeConcini of Arizona, and Donald W. Riegle Jr. of Michigan, as well as Republican John McCain of Arizona—faced charges of improper behavior in the savings and loan meltdown, and they ultimately became known as the Keating Five. All five had met with Keating and a federal regulator about an audit of Lincoln on April 9, 1987. When it became clear in the meeting that Keating might face criminal charges, Glenn had left immediately. Unfortunately, Glenn also shared a lunch with Keating and Speaker of the

House Jim Wright in January 1988, when he believed that no criminal charges would be filed.

Each member of the Keating Five had received financial campaign support from Keating's organization. Glenn garnered $34,000 directly from Keating and friends, and an additional $200,000 went to a Glenn political action committee. (This represented a small contribution compared to what Cranston received, which was close to $1 million.)[79] "After a lifetime of adhering to scrupulous ethical standards, I found myself one of the Keating Five," Glenn wrote in his memoir. "The Republican Party in Ohio asked for a Senate Ethics Committee investigation, and soon afterward, Common Cause joined in the request."[80]

In the summer of 1990, the Ethics Committee's outside counsel Robert Bennett recommended that McCain and Glenn be dropped from the investigation, saying he found no connection between the money they received and any intervention on behalf of Keating.[81] However, the committee decided to investigate all five, apparently because McCain was the only Republican and dropping charges against him would have made Democrats alone appear suspect. About six months later, the panel found that Glenn was guilty of nothing worse than exercising "poor judgment" by having lunch with Wright and Keating.

That was it, but the probe had stained his reputation. During the investigation, Glenn aides saw flashes of his fighter pilot persona when the senator glared at committee chair Howell Heflin. "Once you've got his back up, you're done. He's going to get you. He could focus and did not tolerate objection," sometimes directly confronting those he thought were wrongly accusing him, said Grimes.[82]

On the other side of the world, President Saddam Hussein sent Iraqi troops into Kuwait on August 2, 1990, in a move to capture Kuwaiti oil wells and erase Iraq's large debt to the smaller nation. Iraq then officially annexed Kuwait. Bush immediately expressed US objections and urged Iraqi withdrawal. The UN Security Council ordered an end to worldwide trade with Iraq. Without any retreat by the Iraqis, the Security Council voted November 29, authorizing "all means necessary" to force Iraqis to evacuate Kuwait.

Based on what he had heard in testimony before the Senate Armed Services Committee, Glenn urged Bush to show patience and allow time for trade sanctions to work. He feared another Vietnam and said, "It is absolutely necessary that the American people understand and support any action that this nation takes in the Gulf. This is something we did not achieve in Vietnam."[83]

Calls for caution were unheeded as Congress approved the use of US forces on January 12, 1991. Glenn was among forty-seven senators who voted no.[84] US air and missile attacks began within a week. Battleships entered the fray early in February. A ground attack started February 24. Iraqis swiftly fled Kuwait City, and Iraqi generals released all POWs almost immediately after a February 28 cease-fire was declared. The war had lasted forty-two days. In its aftermath, a Gallup Poll showed that Bush's approval rating had soared to 89 percent—an astronomical number.[85] In the long run, the spike in popularity did not help Bush; he lost his fight for reelection less than two years later.

In the war's aftermath, Glenn repeatedly pushed for a Pentagon accounting of lessons learned in the Persian Gulf. Less than a month after the fighting had ended, he was among a group of lawmakers visiting the Gulf for a briefing by Commanding General Norman Schwarzkopf. The general said the United States needed to be prepared to provide a better strategic sealift; there were not enough US ships available, so he had to deploy foreign ships. He noted that reconnaissance target photos were not good enough and reported maritime minesweeping deficiencies.[86]

Glenn later set up hearings in the Subcommittee on Military Readiness and Defense Infrastructure, which he chaired. In his opening statement, Glenn ticked off a long list of failures in what had been a surprisingly easy victory. Among them were a high percentage of friendly fire deaths, difficulty clearing mines, lack of all-weather tactical reconnaissance, problems getting intelligence information to tactical commanders, inaccurate estimates of damage caused by US weapons, and difficulties in moving equipment and supplies into and within the theater of war. He also pointed out that this war offered the rare advantage of six months in lead-up time, a luxury in times of crisis.[87]

For the first time since his election in 1974, Glenn faced a difficult challenge in the 1992 election. The Keating Five scandal and repeated reports about his ongoing 1984 presidential campaign debt had made him vulnerable. Although Glenn previously had seemed to fit the "Boy Scout" sobriquet, these events had opened the door to Republican allegations that portrayed him as just another dirty politician rather than someone who held himself to a higher standard.

Glenn's GOP opponent was Lieutenant Governor Michael DeWine. In announcing his candidacy, DeWine said, "I can't recall a time when our nation and our people have called out so powerfully for change." Electing a three-term senator for a fourth term was antithetical to that idea, DeWine said, and he promised to run for no more than two terms.[88] Ignoring his own eight years of service in the House of Representatives, DeWine tried to portray himself as a Washington outsider and to position Glenn as a shady insider.

Glenn wanted to avoid a dirty battle if possible. He had waged fierce, combative campaigns when necessary, but as a man in his seventies, he did not fancy the idea of rolling through the gutter with his opponent. He wrote "John Glenn's Clean Campaign Pledge," which he planned to follow and hoped DeWine would honor,[89] but the lieutenant governor aired anti-Glenn TV spots characterized by the *Washington Post* as "brutal."[90] One Glenn campaign document was labeled "Glenn Has No Accomplishments (Or: 'What Are Your Major Accomplishments?')" What followed was a three-page list of what Glenn had achieved during three terms in the Senate.[91]

Campaign staffer John Haseley often rode around Ohio with the Glenns during that campaign. Rather than having a driver, Glenn drove himself, with Annie beside him and Haseley in the back seat reading newspapers.[92] When the votes were counted in November, Glenn won the support of 58 percent of male voters and 60 percent of female voters.[93] Glenn became the first Ohioan elected to four consecutive Senate terms. (DeWine was elected to the Senate in 1994, and in contradiction to his earlier promises to seek no more than two terms, he ran for a third term and lost. He subsequently served as governor beginning in 2019.)

During the 1990s, Glenn devoted a great deal of his time and energy to improving government efficiency. The effort's biggest success was the Chief Financial Officers Act, approved in 1990. Glenn spearheaded the drive for this legislation, which gave the Office of Management and Budget "broad, new authority and responsibility for directing federal financial management, modernizing the government's financial management systems and strengthening financial reporting."[94] The act required that larger executive departments have chief financial officers and that these officers report to appropriate oversight committees in Congress. In partnership with the Reinventing Government initiative launched by President Bill Clinton and Vice President Al Gore, Glenn promoted a variety of bills to make the bureaucracy work more smoothly. Some, such as the Paperwork Reduction Act of 1995, became law.[95] In addition, Glenn authored the Congressional Accountability Act, which became law in 1995. Under this legislation, Congress became obligated to follow thirteen national laws that it had not previously been required to obey. These laws covered issues like civil rights, workplace safety, and labor conditions.[96]

Glenn was disappointed in 1994 about his party's failure to pass a major health care reform plan developed by First Lady Hillary Rodham Clinton. Some Democrats predicted that it would guarantee defeat in midterm elections, but Glenn argued that they should act while they had control of Congress. "What good is a majority if you can't use it for something?" Glenn asked, and Ron Grimes considered that "a pretty courageous thing."[97] The Republicans won a majority in both houses of Congress in 1994—and the Clinton health care plan died.

After the Republican Party seized control of Congress, Glenn lost his chairmanship of the Government Affairs Committee, and GOP senator Fred Thompson took over. The committee had set out to investigate campaign-financing violations, but in 1997, Thompson narrowed the thrust to Clinton's campaign to be reelected in 1996.[98] Glenn "was very frustrated by just the sheer partisanship," Grimes recalled, and the senator's response "was bitter and not in keeping with Glenn's usual demeanor."[99] Republicans believed Glenn was obstructing a meaningful investigation to defend a Democratic president.[100]

Glenn seldom became involved in what could be called populist issues, but in 1995 he made an exception. When news broke that Art Modell, the owner of the Cleveland Browns, decided to move the NFL team to Baltimore, he "blindsided Cleveland and Ohio," in the words of Haseley. "When this issue broke, the fax machine just erupted like a volcano spitting out faxes from people" who had never previously contacted Glenn's office, he said. In response to angry fans, the senator proposed the Sports Fans Rights Act, which would have forced team owners to involve fans in the process of making such decisions and would have given the NFL a partial exemption from antitrust laws, thus enabling it to stop franchise moves from city to city. Although the bill never passed, Glenn triumphed by pushing for the Browns' name and logo to stay in Cleveland and keeping the door open to a new franchise, Haseley said. Rowdy but loyal fans in Cleveland's Dawg Pound went to the Capitol to thank Glenn,[101] and Grimes concluded, "It was like there was a whole new constituency of people, sports fans, that didn't pay attention to what John Glenn was doing or what the Congress was doing" before this issue arose.[102] In 1999 a new iteration of the Cleveland Browns came to life in the city.

Glenn announced in early 1998 that he would not seek reelection in November, when he would be seventy-seven. His sights were set on a higher goal—not higher office but a return to Earth orbit. Clinton took note of Glenn's plans at the annual Gridiron Club Dinner, a closed-door joke fest in a room filled with journalists, politicians, and business leaders. "I see my friend John Glenn sitting over there. Washing down his tube of dessert with a glass of Tang. John, I am proud [of] the mission you will take this year, but I want you to know: If I could compile a list of senators to send into outer space, you, my friend, would be far down on the list."[103]

12

WINNING HIS
WINGS AGAIN

Some dreams never fade. For John Glenn, space's siren song was unbroken. Despite all the years of a full-to-overflowing life, he still had a special thirst for space. Like a marathon runner, eyes wide as he sought a glimpse of the next cup of water, Glenn needed it.

Harnessing the resourcefulness that had served him well as a wartime pilot, an early astronaut, and a determined senator, he found a way to transform his biggest handicap—age—into an asset. The Ohio senator, who planned to retire at the end of his fourth term in 1999, used his perch on the Senate's Special Committee on Aging as a means of snagging a seat on the space shuttle. Glenn's decision to return to space in 1998, at seventy-seven years old, coincided with publication of *The Greatest Generation*, Tom Brokaw's ode to the men and women who fought World War II on the battlefield and the home front. Glenn's desire to return to space demonstrated something else that was special about his generation—the unwillingness to bow to old age as previous generations had. Happily, for Glenn and others, improvements in health care made that stubborn refusal significantly easier to sustain.

During 1995 Senate hearings on NASA's budget, Glenn had asked the agency to send him information on the physiological effects of spaceflight. He received *Space Physiology and Medicine*, a book that tracked more than fifty ways in which spaceflight could affect the human body. "Looking over some of that material," he said, "I noticed that a number of the changes were things that seem to also occur here on Earth as a natural part of the aging process for those in their upper 60s and older." Among them were problems such as bone loss, deterioration of muscles, and sleep disorders. Afterward, he talked to experts at the National Institute on Aging about this phenomenon, and as a result, the institute scheduled three conferences that spawned further research ideas.[1]

Glenn first approached NASA administrator Daniel Goldin in 1995. He argued that similarities between the long-term effects of weightlessness and the symptoms of aging would make it useful to send an older American into space, where new tests could be done. Glenn, of course, had himself in mind as the guinea pig. His hypothesis, says John Charles, project scientist, was "that as aging causes changes in your physiology and even your anatomy that mimic those of spaceflight, then perhaps the aged individual's already adapted to spaceflight and should have a lesser hurdle, an easier adaptation" to life in space.[2] Glenn also made clear that he believed experiments in space might one day benefit the elderly on Earth.

Goldin initially hesitated. NASA had ended civilian flights after schoolteacher Christa McAuliffe's death in the 1986 *Challenger* accident. Before that change in policy, Senator Jake Garn had flown aboard *Discovery* in 1985, and then-Representative, later-Senator Bill Nelson had traveled on *Columbia* in 1986, just before the *Challenger* disaster. Both lawmakers were under fifty-five at the time of their flights. Goldin wasn't sure that Glenn, at seventy-seven, could withstand the stress of space travel. However, astronaut statistics provided some support to Glenn's argument. Since 1959, when NASA had chosen the Mercury astronauts and insisted that all candidates be younger than forty, a lot had changed. The agency subsequently had sent a sixty-one-year-old astronaut, Story Musgrave, into space aboard the shuttle for his sixth flight, and eight people over fifty-five had flown more than once. Astronaut Shannon Lucid had spent six months aboard the Russian space station *Mir* when she was fifty-four.[3]

Glenn first sat down with Goldin and made a formal proposal on June 16, 1996. Goldin made it clear that to fly again, Glenn had to pass rigorous physical exams, and the scientific basis of his proposal had to win the endorsement of scientists and gerontologists. Glenn went to Houston in August 1996 for his annual physical examination.[4] All astronauts who have flown are encouraged to take part in yearly physicals. Consequently, every year since he joined the space agency in 1959, Glenn had undergone a yearly NASA physical—and that made him an appealing candidate. A wealth of data about his health over almost forty years was readily available.[5]

When Glenn saw the doctors, he asked them to go beyond routine testing. They gave him every cardiac test available, including an invasive angiogram. One of the doctors wrote to NASA headquarters that Glenn had passed more physical testing than any astronaut before him.[6] And still Glenn waited for a shuttle assignment. Surely his long, fruitless wait for an assignment after *Friendship 7*'s 1962 flight must have played a role in his thoughts. He reported talking with NASA officials about fifty times between his first conversation with Goldin and winning final approval to join a shuttle crew. Glenn was not entirely quiet and coy while he awaited a decision. During NASA's deliberations, he told the National Space Society, "If given the opportunity, I would most definitely want to fly again on the shuttle."[7]

He was in a meeting in his Senate offices on January 15, 1998, when an aide interrupted, saying he had a call that he might want to take. "You're the most persistent man I've ever met," Goldin said. "You've passed all the physicals, the science is good, and we've called a news conference tomorrow to announce that John Glenn's going back into space."[8]

As he made the announcement, Goldin said:

When somebody comes to you and asks, 'I'm willing to take the risk of spaceflight and serve my country again because I think we can do more to benefit the lives of older Americans. Can I go?' the answer is yes. What an incredible day for John Glenn, for Ohio, for NASA, but most of all, what a day for our country. Because the man who almost 36 years ago climbed into the *Friendship 7* and showed the boundless promise for a new generation, is now poised to show the world that senior citizens have the right stuff."[9]

He revealed that Glenn had been assigned to join shuttle mission STS-95, which was scheduled to lift off aboard the shuttle *Discovery* in October.

Glenn, who sat alongside a giant poster of his forty-year-old self, was surprised by the size of the news conference. "In fact, I was a little bit shocked to the extent of the interest being shown by the press." After twenty-three years as a low-key senator with special interest in the driest of subjects, Glenn had forgotten the reaction he attracted as a risk-taker. After rumors of Glenn's upcoming flight were confirmed, Joe Dirck of the Cleveland *Plain Dealer* wrote, "It is certainly no knock on Ohio's senior senator to note that he is not exactly a natural-born politician. He was always better at his first job. Being an American hero."[10] Thus, American Hero, chapter 2 began.

"I see this as another adventure into the unknown," Glenn told reporters.[11] He said that he would be representing thirty-five million Americans over sixty-five. "On behalf of everybody my age and older, and those who are about to be our age before too many years have gone, I can guarantee you I'll give it my very best shot," he said. Glenn's ambitions went beyond scientific tests. He wanted to make Americans look at what he achieved at seventy-seven and rethink any limitations they expected to accompany aging. Glenn believed that he would be the first of many elderly Americans to take space voyages, thus creating a viable testing database that might help to "lessen the frailties of old age that plague so many people."[12]

After the announcement, Glenn said, "Who would have thought

that age would be an advantage in getting back into space?" The first time he soared into orbit, the United States was at the height of the Cold War–driven space race with the Soviet Union. "The race this time is with age," one Glenn interviewer wrote.

He admitted that Annie, who was seventy-eight herself, "was not wildly enthusiastic" about his plan to return to space. Their two grown children shared her misgivings when they first learned around Christmas 1997 that there was a real possibility of a new spaceflight for their father. Nonetheless, all three eventually made peace with his decision.[13]

Glenn got some teasing from Mercury colleague Alan Shepard, who had beaten Glenn by becoming the first American in space and the only Mercury astronaut on the moon. "John, I'm glad that you're going to give me one more flight for my tax dollars!" said Shepard. He added that he had always thought it was a shame that all of Glenn's astronaut training had been invested in a single flight.[14]

"Because he was still a hero, everybody was sort of really happy for him that he got a chance to go," said Michael Neufeld. "Or almost everybody was really happy for him." Some cynics tried to cast a pall over Glenn's mission by saying it was nothing more than a NASA ploy to generate publicity or political payback for his defense of President Bill Clinton and Vice President Al Gore in 1997 Senate hearings on their 1996 reelection campaign. True, NASA badly wanted a glowing success story; the most famous spaceflights since the initial moon landing had been one near-disaster, Apollo 13, and one all too real space catastrophe—the loss of *Challenger*. "Is it just science?" Goldin said. "No. Inspiration is part of the American psyche,"[15] and the public reaction to Glenn's flight supported that argument. Goldin declared that the decision to let Glenn fly had been entirely his own; he had not consulted the White House.[16]

Although still serving his final year as a senator, Glenn quickly jumped into NASA training. He went to the Kennedy Space Center for the January 22 launch of the shuttle *Endeavour*, where he was greeted warmly by agency employees. Steve Oswald, who had spent a total of thirty-three days in space during three spaceflights, accompanied him up the gantry to the spot where astronauts boarded the shuttle before liftoff. He also took Glenn aboard the shuttle *Columbia*, which was undergoing work in the Orbiter Processing Facility, and SPACEHAB, a unit kept in a shuttle's payload area to accommodate experimental equipment. Then an exuberant Glenn watched the *Endeavour*'s spectacular night launch from about three miles away.[17]

A week later, Clinton's State of the Union Address contributed to the building breathless anticipation of Glenn's flight; in his remarks, the president said, "A true American hero, a veteran pilot of 149 combat missions and one five-hour spaceflight that changed the world will return to the heavens. Godspeed, John Glenn. John, you will carry with you America's hopes and on your uniform, once again, you will carry America's flag marking the unbroken connection between the deeds of America's past and the daring of America's future."[18]

Glenn, who tried not to ignore his Senate duties, had his first day of official training in Houston on a federal holiday, February 16, President's Day. The mission's commander, air force Lieutenant Colonel Curt Brown met Glenn at Houston's Hobby Airport[19] and took him to see the crew's office, which was filled with seven desks—one for each of the mission astronauts. Already, manuals and schedules were in place for study. The next morning began with the weekly meeting for all astronauts, where he was introduced. He asked his new colleagues to call him John, not Senator.

His next chore was a fitting of his launch and reentry pressure suit. When he flew in 1962, he had his own personally fitted suit that matched every detail of his body. By 1998, NASA had multiple sizes available for astronauts. He was fitted while wearing a diaper—something all astronauts donned for shuttle liftoffs and landings. Over the diaper, he wore long underwear topped by a coolant garment, and the suit itself. On his back, he would carry a survival pack and parachute harness. Then he underwent orientation on the crew areas of the shuttle.[20]

The initial plan was to have all crew members sleep in hammocks they could string wherever they wanted, but after further consideration, it was decided that Glenn and Japan's Chiaki Mukai, who were undergoing sleep studies, would sleep in hammocks strung in

John Glenn in the NASA uniform he wore into space in 1998 aboard the shuttle *Discovery*. *National Aeronautics and Space Administration*

boxes that looked like coffins. The rest of the crew could hang their hammocks anywhere from the ceiling to the floor.[21] For some of the sleep exams, Glenn and Mukai had to wear electrodes on their bodies. Mukai could personally position all electrodes except the four on the top of her head, which Glenn would have to place for her. Glenn's bald head was perfect for the positioning of electrodes, but working through Mukai's thick hair was a challenge. He had difficulty finding his way through the hair to her skin. "Chiaki is getting her hair cut shorter for one thing," Glenn said after several frustrating efforts to get through her dense hair. "I don't know whether that will help or not. . . . But I know that I have to practice on the top of Chiaki's head some to make sure that I get that kind of contact." When they were aboard *Discovery*, Glenn and Mukai also would take tests on a computer to record their reaction times and determine whether sleep disturbances affected their cognitive skills.[22]

During training and preflight physical exams, Glenn had at least one uncomfortable day. Members of the crew were shirtless, and he "was a little bit saggy as 77-year-old men are. . . . He was a little bit sensitive," John Charles said, noting that it was a mixed-gender crew. "You know, at one point, he started to try to cover up. . . . We put electrodes on his chest, and he says, 'Oh, hell,' and just let it all hang out."[23] Besides the medical preparations, one of Glenn's training tasks was learning how to use a laptop computer. "It was such a busy mission. We could not afford to have somebody who did half-tasks," payload commander Steve Robinson explained.[24]

Another crucial training topic was the use of what passed as bathroom facilities. There was a small seat, and each crew member had a funnel and his or her own siphon hose that collected urine. Weightlessness carried the threat of urine drops floating through the shuttle. As a result, there were important rules about cleaning the funnel after each use. The seat's handles held the user on the seat, and for defecation, an air-blowing machine drew the fecal matter into waste storage. To keep everything sanitary, astronauts wore gloves throughout the process.[25]

A great deal of training time was devoted to escape routes out of the shuttle. Glenn took part in water escape exercises that simulated a parachute landing, forced him to rappel down the side of the craft, and required him to fall nine to ten feet into a pool of water while wearing all of his heavy equipment. Especially after the loss of *Challenger*, NASA felt obliged to identify every possible means of safeguarding the shuttle astronauts' lives in case of trouble.[26]

One day, he went to Brooks Air Force Base in San Antonio to participate in centrifuge training. He was outraged that facility managers had said they would have an ambulance ready in case he had a medical emergency. The training simulated the equivalent of 3Gs—three times normal gravity. As a Mercury astronaut, Glenn had experienced simulations of up to 16Gs, and as he had expected, he had no trouble this time.[27] Age had landed a spot on the shuttle for Glenn, but he was quite insulted by any suggestion that being old might cause him to require assistance. NASA doctors expressed concern that the dramatic change in gravity at the end of the flight might necessitate carrying Glenn off *Discovery* on a stretcher. To test the possibility that the landing might incapacitate him, they put Glenn on a table that was tilted rapidly. Glenn was quite pleased with the results, noting, "I came out apparently better than many of the younger people do on that particular test."[28]

On the thirty-sixth anniversary of Glenn's February 20, 1962, flight, NASA arranged a press conference for the entire crew—commander Brown; pilot Steven W. Lindsey; mission specialists Scott E. Parazynski, Robinson, and Pedro Duque of Spain; and payload specialists Mukai and Glenn. The senator and lowest-ranking member of the crew was embarrassed by the amount of attention he received. "Most of the questions were of me," he said. "I kept trying to farm them off to the other people, and that didn't seem to work too well."[29] When asked to compare this flight to his one-man 1962 voyage, Glenn joked, "Well, the commander of that flight was a regular guy, and I really enjoyed working with him."[30]

In further training sessions, Glenn learned about photography from space, which interested him greatly,[31] and in sessions with a geologist and a geographer, he received guidance about what kinds of photos were most needed. He underwent regular physical training, some of which he was able to duplicate when he returned to Washington. He lifted weights, stretched, did fast-walking, and rode a stationary bicycle just like the one that would be aboard *Discovery*. He also, of course, received briefings about the medical tests that would be performed on him during the flight.

In March, he took a break from both astronaut training and lawmaking to attend an unusual event at Radio City Music Hall in New York. It was a celebration of *Time*'s seventy-fifth anniversary. Anyone who had appeared on the magazine's cover was invited. It was an eclectic group that included President Clinton, former Soviet leader

Mikhail Gorbachev, actor Tom Cruise, journalist Walter Cronkite, boxer Muhammad Ali, actress Mary Tyler Moore, evangelist Billy Graham, and of course, Glenn. Actor Tom Hanks provided a special and unexpected salute to the astronaut-senator.[32]

NASA bent the rules a bit and allowed Annie to attend training sessions. Annie's presence was unusual but not a problem. Robinson said, "John and Annie Glenn's bond: It was more than—I mean you couldn't call it a relationship; it was an intertwined, interleaved existence, and together, they were more than two people."[33]

Annie attended some photography classes, watched escape drills, and got a chance to see the gym. She especially liked taking part in sampling some of the different types of food and drink available for shuttle flights. Though Glenn recalled only forty-two options, a 1995 NASA article discussed a broad shuttle menu of more than seventy food items and twenty beverages. On the shuttle, each astronaut chose his or her own food in advance. The menu on Glenn's flight included hamburgers, mushroom soup, smoked turkey, beef stroganoff, grilled chicken, teriyaki chicken, ham, spaghetti with meat sauce, tomatoes and eggplant, peach ambrosia, shrimp cocktail, cookies, granola bars, chocolate instant breakfast, juices, lemonade, and many other alternatives. Each person had a dedicated space for his or her chosen meals, plus there was a pantry for snack items, which were fresh. The entrees were dehydrated and required an infusion of water to reach optimal tastiness.[34] The goal was for each astronaut to consume twenty-seven hundred calories a day.[35] This was a far cry from eating applesauce from a tube, as Glenn had done in 1962. As spaceflights became longer, NASA found ways to offer more palatable choices and learned that astronauts could eat food with a spoon because surface tension helped to hold it in place. With the availability of hot water and equipment to heat and cool food, astronauts no longer faced a primitive diet.[36]

During training, the Glenns bridged generations of NASA life. One night, they took the whole crew out for dinner, and on another occasion, they ate at the home of the man who had overseen Glenn's Mercury flight, Chris Kraft.

During eight months of training, Glenn learned details unknown to the public:

- Most astronauts wore two watches, one on GMT (Greenwich Mean Time)[37] and one that showed how much of the mission had elapsed and counted down the minutes until the mission's end.
- During the flight, the astronauts wore casual clothes bought from a catalog store, and they could request whatever they wanted. Glenn had heard that some astronauts were cold on the shuttle, so he asked for a sweat suit with a hood.
- Brown had strict rules of decorum. He demanded silence before liftoff. Except the commander and the pilot, no one should speak after the countdown reached its nine-minute automatic hold.[38]

While Glenn trained, members of the Ohio congressional delegation urged General Mills to put his wrinkled face on a box of Wheaties. Glenn had turned down the chance to become a Wheaties cover boy in 1962, rejecting what ultimately may have been a multimillion-dollar payday.[39] Acknowledging that all previous Wheaties boxes had celebrated athletic prowess, his congressional colleagues argued that he deserved the honor because his life was "filled with service, glory and heroism." This time General Mills was not interested.[40]

Attention to Glenn's role on the flight may have been a distraction to his crewmates, but his presence allowed them to enjoy unique experiences, such as playing host to the president of the United States. Clinton went to Houston in May for a speech about race, and he stopped off at the Johnson Space Center to see the senator. In advance of the visit, Glenn had helped Brown decide what they should show Clinton. While taking a tour of the orbiter mockup and other facilities at the space center, Annie learned from Hillary Clinton, then First Lady, that she had once dreamed of becoming an astronaut.[41]

The National Air and Space Museum invited people from around the world to join the celebration by sending an electronic postcard to Glenn. The museum later printed some of the messages. One person wrote, "Thank you so much for the reminder that the only limits in this life are the ones you impose upon yourself—that with hard work and a little luck, anything is possible." A child told Glenn, "I just wanted to say you're my hero. My grandpa helped build a spaceship. I have to go now. Love." And from around the world he would soon orbit, he received a simple message: "Godspeed from the men and women of Special Operations Command, Central Deployed at Doha, Qatar."[42]

Glenn, who commuted between Washington and Houston, was startled on one trip when he saw red-white-and-blue signs temporarily designating NASA 1, the highway between Interstate 45 and the Space Center, as "John Glenn Parkway." In his memoir, he said, "I thought things had definitely gotten out of hand."[43]

After long months of training, the big day arrived on October 29, 1998. Over the years, Americans had become blasé about shuttle flights. Despite the horrendous sight of *Challenger*'s destruction, most Americans had given up learning astronauts' names after dozens had participated in the shuttle program's ninety-four flights to date. Launches no longer attracted huge crowds. The broadcast networks rarely offered live coverage of shuttle launches.

This flight was totally different. About 250,000 people gathered in Florida to watch the liftoff at the Kennedy Space Center. Among those present were twenty-five hundred journalists. The other three surviving Mercury astronauts—Wally Schirra, Scott Carpenter, and Gordon Cooper—stood in the midday sun to watch the oldest Mercury astronaut again do something unprecedented.[44] (Shepard had died between the announcement of Glenn's flight and the mission itself.) "The first time I saw a large number of people at the Cape since Apollo 11 was STS-95 with John Glenn," Schirra said later. "Now that means that the human being has some factor in this thing to get the excitement up."[45] Glenn's friends from throughout his life showed up to watch the liftoff. Rudy deLeon chatted with an elderly man in a wheelchair while he was awaiting the launch and was surprised when someone later asked him what he and Ted Williams had been discussing.[46] Expectations of a much bigger than usual launch crowd led NASA to spend $204,554 on security.[47]

"The scenes of people jamming Times Square to see live TV

coverage of the launch, and thousands lined up on beaches in Florida for the liftoff, reminded me of those days [of the Mercury program], which don't seem so long ago, when there was no bigger story," wrote Cooper, who had talked to Glenn from Australia during his 1962 flight and explained the glowing lights of Perth.[48]

Some of those in the crowd had seen Glenn's 1962 launch, and many stopped by the US Astronaut Hall of Fame near the launch site and left messages for Glenn and his crewmates. A surprisingly high number of those messages came from non-American tourists. A man from Wales wrote, "Godspeed. Best of Luck Glenn," and a twelve-year-old girl from Halifax, England, told him, "Good luck. Wish you all the best on your 2nd trip." A Swede inscribed the message book, saying, "The future belongs to those who believe in their dreams." An American man told Glenn, "I was watching in 1962 and saw your triumphant parade in DC, where I lived. We came here from Dublin, Ohio, to see your historic return. Something I and my children will remember forever. God bless." Another noted that Glenn "made me proud to be an American."[49]

While the crowd awaited liftoff, Kennedy Space Center technicians closed the hatch at 12:30 PM, just as Air Force One delivered the Clintons to the shuttles' landing runway. The countdown proceeded without delay until the usual T-9 minutes hold. A master alarm that sounded during cabin leak checks caused another 8.5-minute delay before the countdown began again. The range safety officer called for an additional hold at T-5 minutes in response to a small private plane that had strayed into restricted air space. Then the countdown sped along to liftoff at 2:14 PM. After the start of the main engine but before booster ignition, the drag chute compartment door fell off. This small accident would not affect the mission; managers later decided against trying to use the drag chute during landing. Moments before liftoff, Scott Carpenter, who had so memorably said, "Godspeed, John Glenn," in 1962, gave the shuttle crew a similar benediction. He stood in an NBC booth and broadcast a message: "Good luck to the commander and crew of the Shuttle and, once again, Godspeed, John Glenn."[50]

When it finally happened, the launch was a much bigger and louder jolt than Glenn had experienced on *Friendship 7*. Mercury astronauts were not always certain whether liftoff had occurred until they saw the capsule's clock starting. Liftoff thrust in 1962 had been 360,000 pounds, while it was 7 million pounds in 1998. The weight of his first vehicle at liftoff had been 4,256 pounds as opposed to 4.5 million pounds on his second flight.[51] *Friendship 7* had been on its first and only mission, while *Discovery* was setting out on its twenty-fifth of thirty-nine voyages into space.[52] An internal shuttle video showed Robinson and Glenn celebrating the excitement of liftoff by jubilantly high-fiving and shaking hands.[53]

At liftoff and landing, Mukai, Robinson, and Glenn were not on the flight deck with the beautiful view. "We were down in steerage," the senator joked. "It was us and a couple of toadfish."[54] However, during the flight, he had plenty of time to enjoy the view. Residents of Perth, Australia, kept their lights on for Glenn as they had in 1962, and he could see it all over again—two bright squares of light in the dark of night. "It was a long time ago that I looked at the same thing and it looks beautiful. . . . I think Perth has grown a little since the last time."[55] On another occasion, when Glenn looked down at the planet Earth, an image he had never expected to see again with his own eyes, a tear formed in his eye and remained there. "In zero gravity a tear doesn't roll down your cheek," he explained later. "It just sits there on your eyeball until it evaporates."[56]

Two days after the launch, the *Houston Chronicle* stepped forward to belittle those who had criticized the flight: "For one day, after a far-too-long 36-year wait, this country stopped, just as in 1962. We had a common uplifting bond—if even for a few fleeting minutes—before returning to cell phones, faxes, and pagers."[57] Tom Shales of the *Washington Post* wrote that Glenn's return to space was lifting American spirits after the House impeached Clinton and the Senate prepared to try him on charges of perjury and obstruction of justice in his attempt to hide an illicit relationship with White House intern Monica Lewinsky. Shales quoted author Isaac Asimov, saying that on that day Americans "had reason again to believe 'there is a future . . . and the future will be good'—if only we can get through the next few months. . . . Most of the glory goes to Glenn, but there's some left over for those of us who only watched and wept."[58] As on his first flight, Glenn did not fully understand the emotional chord his daring struck among the American public. He was too absorbed in the flight itself.

Shortly after leaving Earth's gravity, Glenn quoted his own words from 1962, telling mission control, "Zero G, and I feel fine." When Glenn had last used those words to describe weightlessness, he had been a weightless Mercury astronaut trapped in barely thirty-six cubic feet of space. On the shuttle, 332 cubic feet of breathing room allowed him to float from floor to ceiling, deck to deck, and throughout *Discovery*. "Let the record show John has a smiling face, and it goes from one ear to the other one," Brown reported.[59]

Glenn enjoyed his interactions with his crewmates, a diverse group of space travelers. Brown, a veteran of four spaceflights by 1998, commanded three, including one that followed Glenn's flight. He was an avid sportsman from North Carolina. For Lindsey, a native Californian, STS-95 was his second shuttle mission. He went on to command three more. Both Brown and Lindsey were graduates of the US Air Force Academy and spent part of their military careers as test pilots, as Glenn had.

Parazynski, known as a joker, attended junior high in Dakar, Senegal, and Beirut. In high school, he studied at the Tehran American School in Iran and the American Community School in Athens. He graduated from Stanford Medical School and was among the top ten luge athletes in the 1988 US Olympic Trials. After this flight, which was his second, he flew twice more, and in 2009 he became the first astronaut to reach the summit of Mount Everest. In his memoir, *The Sky Below: A True Story of Summits, Space and Speed*, Parazynski, who was a toddler at the time of Glenn's first flight, wrote, "No space walks planned, which is a disappointment for me after my first taste of EVA, but having a living legend like John aboard more than compensates." He titled his STS-95 chapter "Physician to the Stars."[60]

Mukai, also a physician, was educated in her homeland, Japan, and became a science astronaut for NASDA, the National Space

Development Agency. She was the first Japanese woman to travel in space and the first Japanese person to fly twice. STS-95 was her second and final flight.

Robinson, who started flying at the age of fourteen, worked as a graphic artist, surveyor, musician, and radio disc jockey before joining NASA's Ames Research Center as a research scientist in 1979. He earned his doctorate in mechanical engineering from Stanford University in 1990 and continued working as a graphic artist during his career as an astronaut.

Duque was educated in Spain. A lover of water sports, he had trained to become an expert on Russia's Soyuz-TMA spacecraft and went into orbit aboard *Discovery* five years before he had an International Space Station mission on the Russian spacecraft.

Discovery's crew planned to conduct more than eighty experiments. The most complicated shuttle activity was deployment and retrieval of *Spartan*, a satellite monitoring solar winds. Robinson used the shuttle's fifty-foot-long robot arm built in Canada to maneuver *Spartan*, deploying the satellite on November 1 and retrieving it two days later.[61] During most of that period, *Discovery* remained about thirty miles from *Spartan*. The crew also deployed the Petite Amateur Naval Satellite.[62] The project was initiated as a means for students at the Naval Postgraduate School Space Systems Engineering and Space Systems Operations to design and operate space hardware. More than fifty master's theses focused on the device between 1989 and its deployment. Once deployed, it provided telemetry that students at the school monitored.[63]

The mission's medical goals were "to determine how our understanding of the aging process can be improved in the absence of gravity; to correlate aging research to a better understanding of physiological deconditioning in space; to improve understanding of the basic mechanisms of aging; and to develop countermeasures where possible." NASA and the National Institute for Aging planned experiments to study osteoporosis, balance disorders, sleep disturbances, cardiovascular deconditioning, immune response, protein turnover's effect on muscle volume, postflight recovery of equilibrium, cardiovascular responses to standing before and after flight, and impaired body temperature regulation. When weightlessness causes these effects among astronauts, their bodies gradually restore themselves after returning to Earth; however, for the elderly, there is little hope of reversal.[64]

Each morning, the astronauts' eight-hour sleep period ended with a different wake-up song, usually chosen by one of the astronauts. Twice, the crew awakened to music that was special to Glenn: one day, at Annie's request, the astronauts' morning opener was "Moon River" sung by the couple's longtime friend Andy Williams;[65] on another day, it was "Voyage into Space," an original orchestral piece written by another friend, conductor-composer Peter Nero.[66]

The mission had many shuttle-to-Earth communication events, and Glenn participated more than anyone else. He took part in news conferences with reporters from around the world, and special sessions for Japanese and European reporters because the crew included Duque, who represented the European Space Agency, and Mukai from Japan. Walter Cronkite, who joyously left retirement to cover Glenn's flight for CNN, joined NBC's Tom Brokaw, CBS's Dan Rather, ABC's Peter Jennings, and other TV journalists for a question-and-answer period with Brown and Glenn. In that interview, Glenn gave an answer that went to the bedrock of his feelings about the United States—his belief that the need to see what's beyond the next hill or across a raging river is not just what settled the frontier, it is what made the United States a world leader. In answer to a question from Neil Cavuto at Fox News in New York, he turned the conversation into a discussion about his crewmates and how exploration could improve American life. "That's what built the United States, was a curious, questing spirit that tried to go out and learn new things. We're on the cutting edge of science with some 83 projects on this particular flight and these are brilliant people I'm up here associated with, I'm just honored to be here."[67]

On October 31, Brown and Glenn answered questions from students at the Center of Science and Industry in Columbus, Ohio; the Newseum in Roslyn, Virginia; and John Glenn High School in New Concord. When asked whether his return to space was worth the wait, Glenn said, "Yes, it certainly was. Not just as a personal experience, but also in looking forward to some of the research we are going to do that is going to benefit everybody back there on Earth, benefit all of us into the future." Glenn also told the students that the biggest influence in his childhood was his hardworking and entrepreneurial father—a man without much education who helped to shape his son's work ethic and optimistic outlook on life.[68] During his almost nine-day mission, Glenn also appeared live on *The Tonight Show with Jay Leno* and spoke to the Japanese prime minister.

He laughed about all of the medical testing, reporting from orbit that Parazynski had "taken my blood so many times that when I see him, I say, 'Here comes Dracula.'"[69] Ten blood samples and sixteen urine samples were taken from Glenn during the flight. Other members of the crew participated in some medical testing, but Glenn was the pincushion in chief.[70] On some days, he fed bone cell cultures in a test of cell functioning under microgravity conditions.[71] (Parazynski later said that his responsibility for Glenn's health made this the most stressful flight he took. "It was wonderful to fly with John, but what if something really happened to him? I might as well not come back home was the kind of mindset that I had.")[72]

Even when he slept, Glenn's body churned out a variety of test results. On the first night of the flight, the 9:00 PM status report indicated Glenn would swallow a thermistor capsule that would measure his core body temperature.[73] On other nights, he and Mukai wore electrodes to measure brain waves, eye and body movements, and respiration. All of the data was collected on a digital sleep recorder.[74] Glenn also wore electrodes and a holter monitor to record his heart rhythm.[75] Each day, Brown, Robinson, and Glenn filled out a back-pain questionnaire, and Glenn and Mukai kept a record of what they ate. Glenn also helped his crewmates with their projects. For example, he assisted Duque, the first Spanish astronaut, with an astroculture plant-growing experiment and with an antitumor study.[76]

As *Discovery* orbited Earth 134 times, Glenn sent emails to Annie. On one day, Parazynski hijacked an email after it had been addressed. He wrote, "Dear Annie: What do you mean, no third flight? I was already planning my 4th mission, a 6-month stay on the International Space Station. . . . Just kidding! (All the best to you from Scott and the rest of the crew)—here's John." Not a great typist, Glenn followed Parazynski's message with "now how do you like yhat ffor a start?"[77] In another message, Glenn wrote, "Are you really sure you ddon't wwant to be the first old fogies couple aboaard the space station????"[78] He wrote a much neater email to Clinton. Looking at it, one must wonder whether another astronaut took his dictation, or perhaps someone at the White House cleaned it up before it was printed for the president.[79]

Eventually, the astronauts' mission watches neared the end of their time. Following an hour-long descent, *Discovery* glided to a landing on the three-mile-long landing strip at the Kennedy Space Center on November 7 at 12:04 PM EST. A visibly shaky Glenn, who had been indignant at the suggestion that he leave the shuttle by stretcher, managed to join the rest of the crew on the traditional inspection walk around the shuttle after landing. He later told Dale Butland, "I determined that if it killed me, I was going to walk off that thing."[80]

The astronauts all were reunited with their families on that day and spent that night near the Kennedy Space Center before returning to Houston the following day.[81] Weeks later, Glenn admitted that the astronauts' exit from the shuttle was delayed almost two hours by his bout of vomiting as his body readjusted to normal gravity. Fluid within an astronaut's body suddenly moves into the lower part of the body as gravity takes effect, and this often causes light-headedness. For that reason, astronauts typically drink a lot of water before landing. "I preloaded with too much fluid coming down," Glenn explained.[82]

When the crew returned to Houston, Glenn received a hero's welcome. He told the hundreds of people who greeted the crew at Ellington Field that every shuttle mission deserved the kind of attention that his voyage had received. "We have gone all the way from just flying to see if we could do space travel to using it as a laboratory," he told the crowd.[83]

Nine days after the crew's return to Earth, they were celebrated in a New York City ticker-tape parade. This event did not rival Glenn's 1962 parade, which had attracted four million people. However, the city had seen two ticker-tape parades in the previous month—one for the World Series Champion New York Yankees and one for Chicago Cubs home-run hitter Sammy Sosa—so the city may have been suffering from a bit of parade fatigue. Nevertheless, a crowd of 500,000 is worthy of note. Douglas Martin reported in the *New York Times*, "There were many cheerful scenes of people enjoying themselves during their brush with history. Fathers hoisted children on their shoulders, children waved American flags, and people lined up to buy commemorative T-shirts."[84] Cronkite reported that Glenn "was a hero to us then and he is a hero to us now.... His fortitude will benefit all of us, whether we are men of 82 years or children of 3 or 4, Sen. Glenn's journey shows that Americans of every age are vital contributors to our society."[85] (The retired newsman's eighty-second birthday occurred during *Discovery*'s flight.)

Glenn's family was proud of his accomplishment but relieved when the flight was over. Writing about Glenn's son, Dave; his wife, Karen Sagerstrom; and their two teenage sons, one reporter noted, "There is a quiet dignity about the family, and despite the hero's welcome that 77-year-old Grandpa received yesterday in a New York ticker-tape parade, they are profoundly unamazed by the accomplishment. They know as well as anyone what John Glenn is capable of doing. But their pride is weighed down by an earthbound fear, a nagging and constant concern for the safety of a loved one." When he heard that his father would ride a shuttle into space, images of the *Challenger* disaster had haunted Dave Glenn.[86]

After returning to planet Earth, John Glenn and the rest of the *Discovery* crew faced a long list of requests for public appearances, including a ticker-tape parade in Houston, visits to the astronauts' hometowns, trips to NASDA and ESA, events at NASA-related facilities, an appearance by Glenn on *Meet the Press*, and an appearance by Glenn, Brown, and Lindsey on *The Tonight Show with Jay Leno*. Most requests, but not all, received positive responses.[87] In the Houston parade, Glenn rode on a 1926 Rolls-Royce, with Annie at his side. The parade was preceded by a luncheon sponsored by then Texas governor George W. Bush. Glenn said that he appreciated everyone's praise, but he tried to turn the spotlight on his crewmates: "I feel like costume jewelry at Tiffany's compared to these people."[88]

After the flight, Brown explained why he didn't mind the attention given to Glenn. "The fact is, the public doesn't know shuttle crews. People will never know how much fun Scott is. They'll never know the infectious enthusiasm Chiaki brings to her work. But they know John. And if he brings people closer to the space program and generates enthusiasm for it, that's good for all of us."[89]

Glenn's return to space did exactly what Brown described: it encouraged Americans to think about space travel and to see the space program as something that was oriented toward improving life on Earth—not just crossing the next frontier. The scientific studies in which Glenn participated did not unearth any major medical revelations. Instead, an ironic twist emerged: despite a dramatic age difference between Glenn and the other subjects, the oldest of whom was nine years old when Glenn first orbited Earth in 1962, his readings were not markedly different from those of his crewmates, according to Charles. After receiving the results of postflight DEXA scans, Charles told Glenn, "Your whole-body bone density, body composition, and percent body fat was comparable to that of the other two STS-95 subjects."[90]

And so it went from test to test. Glenn did not have significant additional muscle atrophy, but the rest of the crew showed slight increases.[91] His balance issues registered few differences with his fellow travelers.[92] Cardiovascular tests also were similar.[93] Glenn's blood tests did indicate higher in-flight stress levels than his crewmates' readings. That could threaten the immune system of an elderly astronaut but didn't add significantly to knowledge about aging or the effects of space travel.[94] A report on sleep showed that Glenn's own daily sleep self-estimates indicated insomnia during the flight. Nevertheless, at the point in the typed report where his insomnia is described, there is a handwritten note that simply says, "Glenn denies."[95] During communications with people on the ground, Glenn talked about how much easier sleep was with the ability to avoid pressure points in a weightless environment.[96] (It is worth recalling that this same John Glenn also did not admit to "stomach uneasiness" after bobbing in the waves after his Mercury splashdown; instead, he allowed doctors to record only that he had "stomach awareness.") The bottom line on Glenn's hypothesis was that aging "did not provide any benefit in terms of pre-adaptation to weightlessness," Charles said.[97] Even if Glenn's findings had been extremely different from those of his crewmates, he represented a sample of one. NASA has no plans to send additional elderly astronauts into space, so his results are likely to stand alone for the foreseeable future. Therefore, it will remain impossible to make generalizations based on those findings.

No matter what the test results were, Glenn had turned the world upside down. "The old hero! What an idea!" Robert Rosenblatt wrote in the August 17, 1998, issue of *Time*. "Not an older person who was formerly great, like a DeGaulle or a DiMaggio or a Golda Meir, but an active hero in old age. Fiction gives us such characters from time to time, but reality is too real. The flesh-and-blood old hero may need a special logic to return to the field of conquest. Try this on for size: So much of space exists in the past (dead stars shine). Why shouldn't someone be able to retrieve the past in the future and be young again by being old?"[98] Here, in a nutshell is the story of John Glenn's reinstatement in the space program. In a nation that glorifies youth, he transformed the common definition of a hero.

On this Glenn voyage, there were fewer unknowns than during his 1962 mission. Though he was older and the flight was longer, the individual pressure to perform was lighter. He was not alone in a tin can where little body movement was possible. He could float from place to place and had time to enjoy the ride. But he stands alone as the oldest person on Earth to embrace such a challenge.

EPILOGUE

Inspiration is what a hero provides, and John Glenn's second space-flight, like his first, delivered hope, happiness, and a sense that Americans could be greater than they are. Again, he gave the nation a shared feel-good moment. In the process, he found a new brigade of avid followers—new believers—by touching the lives of late twentieth-century children and motivating members of his own generation to pursue unrealized and abandoned dreams.

"Boy, you're changing my life," a seventy-four-year-old man told Glenn at the Houston Airport. The man joyously explained that he had always wanted to climb Mount Kilimanjaro. "Now I'm gonna do it!" he declared, saying Glenn's decision to go into space at seventy-seven had encouraged him. Afterward Glenn said, "So I may have killed a man on Kilimanjaro for all I know. . . . I've noticed that maybe because of all this, people are seeing themselves differently. . . . They're realizing that older people have the same ambitions, hopes, and dreams as anyone else. I say you should live life based on how you feel and not by the calendar."[1]

In the wake of the announcement of Glenn's second flight, peo-ple all over the country would muse: "Well, if John Glenn can fly in space at age 77, why then, I can—fill in the blank," said astronaut

Steve Robinson.[2] And again, Glenn offered good news to the baby boomers who had cheered him as children and now learned from his example that as they approached senior citizen status, they did not have to accept a sedentary life. At the same time, he helped them face life's realities. In the speech announcing his Senate retirement, Glenn conceded that despite all of the advances in medicine, "one immutable fact remains. There is still no cure for the common birthday."[3]

Glenn retired, but he had no intention of spending the coming years in a slow-moving rocking chair situated on the outskirts of public life. After leaving behind both the Senate and spaceflight, he and Annie split their time between their home in Potomac, Maryland, and Ohio. He was eager to get to work at the John Glenn College of Public Affairs at Ohio State University, later known simply as the Glenn College. He wanted to offer education to undergraduate and graduate students in a variety of fields that would train them for roles in public service, including nonprofit activism, world affairs, scientific policy, leadership, and policymaking. To add new perspectives to the students' education, Glenn became an adjunct professor himself and shared his own wide-ranging experiences. He had a prescription for good citizenship that he outlined in his retirement statement:

> Don't tune out, cop out or drop out.
> Don't give in to complacency and cynicism.
> Don't ignore what is bad but concentrate on building what is good.
> Don't take America and the values reflected in our form of government for granted.[4]

These were good guidelines for Americans headed into the twenty-first century.

In 2000, Glenn found himself honored by a postage stamp for the second time in his life when the US Postal Service issued a stamp to mark his restoration to the ranks of space travelers. A nationwide vote of interested Americans chose the shuttle image as one of fifteen commemorative stamps recalling moments from the 1990s. "John

Glenn's return voyage did more than advance our knowledge, it lifted our spirits," said Deborah Willhite, senior vice president, government relations and public policy of the US Postal Service, who unveiled the stamp.[5]

Though his astronaut days were behind him, Glenn remained fascinated by space and was thrilled one day when his cell phone rang, and the voice on the other end said, "Hi John Glenn, this is Scott [Parazynski] from outer space."[6] He was delighted to hear from Parazynski, the Count Dracula who had drained his blood so often aboard *Discovery*, and he clung to his sense of wonder about the ways rapidly evolving technology triggered new experiences, such as receiving a phone call from a friend in outer space. So many things he had experienced had been impossible to imagine in his youth.

Glenn never ceased to be a fervent supporter of the space program, its employees, and its future. In 2003, when the shuttle *Columbia* burned up on reentry thirty-nine miles above Texas and moving at 12,500 miles per hour, NASA lost all seven astronauts and one-fourth of the existing shuttle fleet. An investigation revealed the cause was launch damage to part of the system that protected the shuttles from heat during reentry.

On the day after the accident, Glenn appeared on *Meet the Press*, arguing that this loss should not lead to thoughts of permanently grounding the shuttle fleet. "I think maybe it's good . . . with the way the conversations have gone yesterday and even this morning to remember that we're not up there in space just to joy ride around," he said. "We're up there to do things that are of value to everybody right here on Earth." He noted that *Columbia*'s mission included ninety experiments. Arguing for continued shuttle flights and no major new manned exploration goals, he said, "My attitude has always been that at each step as we go along in the space program, we should maximize the research return as a benefit right back here on Earth."[7]

Glenn received a letter from a first-grade teacher expressing her students' sadness about the loss of *Columbia* and her crew. The class had planned to write Glenn on February 20 to mark the anniversary of his *Friendship 7* flight, but the shuttle disaster on February 1 had preempted that communication. The teacher, nevertheless, expressed the children's gratitude for Glenn's role in space exploration more than forty years after his first flight,[8] a sign that another generation of youngsters recognized Glenn as someone special who had given much to his country.

In July 2008 the eighty-seven-year-old Glenn testified before the House Committee on Science and Technology. He spoke in support of the International Space Station and against President George W. Bush's suggestion that the United States should alter its focus. Glenn felt Bush's plan to send astronauts back to the moon and to Mars while ending any research that did not contribute to those two goals would draw NASA away from the space station immediately after its completion. The United States had asked fifteen other nations to participate in developing and using the station, which had cost at least $150 billion and was constructed in space. Glenn considered it wasteful for the United States to make less than full use of the space station.[9] Ultimately, he was on the winning side in this battle as no new moon or Mars expeditions were scheduled.

A few years later, Bush's successor, Barack Obama, alarmed Glenn by endorsing retirement of the shuttle fleet. Glenn fought the change, saying he feared that eliminating the only viable US spacecraft could limit American use of the International Space Station. In a white paper he posted on the Ohio State University website, he contended that the United States should not abandon its leadership in spaceflight. He recommended flying the shuttle only twice a year at an annual cost of $1.5 billion, which he said would be a small price to pay after contributing $100 billion to the station.[10] He lost this fight, and the shuttle program ended with the STS-135 mission when *Atlantis* landed in Florida on July 21, 2011, forty-two years and a day after Neil Armstrong set foot on the moon. Among 135 shuttle flights, 133 were successful. Two—STS 51-L's *Challenger* mission in January 1986 and STS-107's *Columbia* flight in February 2003—had ended in tragedy.

After "retirement," Glenn remained in demand as a writer of articles about flight in the air and in space. In some cases, he took tough stands on issues; in others, he wrote commemorative pieces, as he did in 2003 to mark the one hundredth anniversary of the Wright brothers' first flight.[11] The famed brothers got their start in Dayton, and as a proud Ohioan, Glenn "could drop a Wright brothers quote at a drop of a hat,"[12] Ron Grimes said. Still in love with flight, Glenn kept his pilot's license through his eighties and gave up flying only when knee weaknesses made it difficult for Annie and him to board a small private plane.

Beyond his continuing role as the best-known and most visible astronaut, Glenn participated in government research projects. He chaired the Clinton administration's National Commission on Mathematics and Science Teaching for the 21st Century. This body, which finished its work as Clinton was preparing to leave the White House in 2000, recommended a major US investment in education to elevate standards in teaching of math and science.[13] In the report's foreword, Glenn wrote, "The future well-being of our nation and people depends not just on how well we educate our children generally, but on how well we educate them in mathematics and science specifically."[14]

As he was about to turn eighty-seven, Glenn joined the Strategic Posture Commission, a panel organized in June 2008 by Congress to identify "the most appropriate strategic posture and most effective nuclear weapons strategy."[15] The commission issued a report to Congress in 2009.[16] It did not consider newly elected President Obama's call for a world free of nuclear weapons. Instead, its conclusions endorsed maintaining much of the status quo with a goal of retaining nuclear weapons parity with Russia and continuing to use three nuclear delivery systems—ground-based missiles, bomber aircraft, and submarine-based weapons.[17]

Despite his continuing involvement in somewhat dry topics, Glenn did not pass up opportunities to have fun. His wife and an assistant convinced him to appear on the much-honored situation comedy *Frasier*.[18] The series, which won more than thirty-five Emmy Awards, had debuted in 1993 and remained on the air until 2004. In a 2001 episode entitled "Docu.drama," the former astronaut played a somewhat fictionalized version of himself. While Dr. Frasier Crane is deeply involved in an argument in an adjoining room, Glenn takes over the microphones in Frasier's radio studio and tells listeners that astronauts saw many UFOs during their travels: "Believe me, we all know—Armstrong, Aldrin, all of us. But the boys upstairs were worried about widespread panic, 'War of the Worlds' type stuff. So we kept our yaps shut, and now we only see it in our nightmares." The preoccupied Frasier never realizes what Glenn has revealed, and the former astronaut manages to retrieve the only copy of his taped confession.[19] As in almost everything he did after leaving the military, Annie was with her husband throughout production, and when cast members autographed a script, it was for "the Glenns," not just John Glenn.[20]

More than once, Glenn served as narrator for concerts by his friend Peter Nero, who was a composer, pianist, and pops conductor. Nero called upon him to serve as a reader for his composition "Voyage into Space." The script's closing lines salute the adventure of spaceflight by explorers such as Glenn: "For spaceflight is the epitome of our scientific and engineering skill, our dedication to the future, our need to understand our world and the universe beyond, and our constant search for answers that will make life on earth better for all mankind."[21]

In 2009 a ghostly issue from Glenn's Mercury astronaut days raised its ugly head. At a time when 30 percent of astronauts were women, he was forced again to disavow any opposition to women in space. Complaining that he had been repeatedly misquoted on this subject, Glenn told an interviewer that it "was fine with me" when NASA changed the requirements for astronauts, making it easier for women to join the corps. He pointed out that at least one woman had commanded a space shuttle mission by then.[22] (As of 2020, women represented 34 percent of the astronaut corps. Two had led shuttle missions, and astronaut Peggy Whitson had commanded the International Space Station for six months on her second tour of duty there. She participated in ten space walks, and her 665 days spent in space gave her the US record for the most time spent in space. Sunita Williams later served as the station's commander.) Glenn did not enjoy being haunted by decades-old statements that were being judged by twenty-first-century standards.

John Glenn receives the Medal of Freedom from President Barack Obama in May 2012. *National Aeronautics and Space Administration*

On his ninetieth birthday in 2011, Glenn received salutes from many friends and acquaintances. "Whether he was flying in the air or floating in space, walking the campaign trails or in this chamber, he remained grounded in the New Concord roots and always by the steady hand and constant love of Annie," fellow Ohioan Sherrod Brown said in remarks on the Senate floor.[23] Senator Harry Reid remembered one day when a double-Dutch skipping rope championship team from his home state of Nevada came to the Hart Building near the Capitol to give him a demonstration. Reid tried unsuccessfully to mimic their moves and "could not get one step." Glenn, then a man of roughly seventy, jumped in, and "he was perfect," Reid conceded with a bit of envy. He praised Glenn's Senate storytelling and admitted, "I just so loved John Glenn."[24]

Even in old age, it seemed Glenn was always on the go. He received the Congressional Gold Medal in November 2011 alongside Neil Armstrong and the rest of the Apollo 11 crew. Via email, Armstrong had sought Glenn's help in polishing his speech for the occasion.[25] Glenn also received a Medal of Freedom in May 2012 from President Obama.[26]

At that time, Glenn and Scott Carpenter were the last living members of the Mercury Seven. In 2013 a stroke and its aftermath claimed Carpenter's life, and Glenn stood alone. At ninety-two, he delivered a eulogy for his one true friend among the original seven astronauts and his one-time next-door neighbor. Glenn recalled meeting Carpenter during the grueling tests that helped to trim a group of thirty-two prospective astronauts to seven. "While the rest of us went through the tests as a means to a hoped-for end, Scott seemed to really enjoy the testing. He was exploring an unknown. It was not only the competition with others, but the competition with himself and what he thought were his own limits."[27] In the mid-1960s, Carpenter had given up space travel to explore life under the sea. He lived underwater for thirty days in association with the navy's SEALAB program, and he worked alongside renowned oceanographer Jacques Cousteau. Glenn said farewell to his friend by celebrating "a notable life, a life of achievement, a life of curiosity and enquiry. The life of our friend, Scott. Godspeed, Scott Carpenter."[28]

Throughout his years as a celebrity, Glenn made many positive and personal contributions to American life away from the public stage. He touched the lives of many Boy Scouts over the years, writing them personal letters congratulating them on their achievements. Furthermore, his life's work continues through the Astronaut Scholarship Foundation created in 1984 by the six then-surviving Mercury astronauts and Betty Grissom, widow of Gus Grissom. The foundation awards more than fifty scholarships per year to science students as well as making grants to university science programs. It honors new generations of astronauts as well by managing the selection of new inductees to the US Astronaut Hall of Fame, which now has more than ninety members.[29]

Even as late as 2014, when Glenn turned ninety-three, his calendar was packed with appointments,[30] but as would be expected for a man of his age, more and more appointments with doctors appeared on his schedule. In July 2016 the former astronaut, who once had a remarkably chiseled and fit body, was embarrassed at the dedication of the John Glenn Columbus International Airport because he didn't want to be seen using a wheelchair as a result of declining eyesight and failing health.[31] Instead, he walked with a cane.[32]

Less than six months later, he died on December 8, 2016, at ninety-five. The cause of death was not specified, although he had heart valve replacement surgery and a stroke in 2014. Just days before his death, Glenn contacted Rudy deLeon to talk about political developments in the transition period leading up to Donald Trump's presidency. Their mutual friend "General [James] Mattis had had an interview with Trump about being the Secretary of Defense," deLeon said, "and so Senator Glenn was very interested in what I thought and what was going on. And so, we ended up doing a three-way call with Secretary Mattis."[33] DeLeon remarked that despite his nearness to death, Glenn was alert and focused on Mattis's potential future.[34]

After Glenn's death, newspaper after newspaper proclaimed that the United States had lost a hero. Senator Brown, who had first met Glenn when he was a sixteen-year-old Boy Scout at a dinner in Mansfield, Ohio, recalled how Glenn had escorted him down the aisle when he was sworn in as a senator. "What made John Glenn a great senator was the same quality that made him a great astronaut and an iconic American hero: he saw enormous untapped potential in the nation he loved, and he had faith that America could overcome any challenge."[35] Ohio's other senator, Rob Portman, who had taught at the Glenn College, wrote on his website that Glenn "was so proud of the college because he believed it was critical to get more young people involved in public service. . . . In fact, the Glenn College may ultimately become his greatest achievement, because through it, his legacy will be to continue to inspire and train thousands of young Ohioans to pursue public service like he did."[36] Senate majority leader Mitch McConnell of Kentucky recalled, "John Glenn said his childhood was like something out of a Norman Rockwell painting, but his life was anything but typical."[37]

Two days after Glenn's death, he was once again being portrayed as a hero in a major film—*Hidden Figures*. In this movie, based on a best-selling book about African American women who labored in obscurity as mathematicians for NASA and its predecessor agency, Glenn is not a leading character, but he is heroic as an astronaut and touchingly as a human being who trusted these women and their skills. As they performed vital work in the early space program, these hidden heroes faced all of the prejudices common in the 1950s and early '60s. They had to accept indignities, such as segregated bathrooms, which were a part of their daily lives in Virginia, where many of NASA's early space-flight management and planning operations were then based. In both the film and the book on which it was based, Glenn paid no attention to the women's skin color. He treated them like so many other people he met in his life, as equals—or more than equals—because he relied on one woman's mathematical skills to verify computer findings that would guide his flight safely through space.[38]

While he lived on movie screens, Glenn's body lay in state in the Ohio State Capitol, and thousands lined up to walk past his casket.[39] Afterward, there was a memorial service. "John defined what it meant to be an American, what we were about, just by how he acted," Vice President Joe Biden said. "It was always about promise. We were a country of possibilities, opportunity, always a belief in tomorrow." Biden continued, speaking directly to Annie Glenn. "Together, you and John taught us that a good life is not built on a single historic act, or multiple acts of heroism, but on a thousand little things, a thousand little things that build character—treating everyone with dignity and respect."[40]

Glenn's children, David and Lyn, spoke about their extraordinary father. "In your heart, Dad," Lyn said, "you remained a small-town boy, and you wore your celebrity lightly and taught us the same. Though you met presidents, CEOs, kings and queens, you had a common touch with people at a gas station or at a Bob Evans, where you and mother would stop for fried mush and scrambled eggs." She recalled that her father "lived many lives in one life, with honesty, grace, belief in our country, and the honor of public service. I am proud and so grateful to say you're just my Dad. Thank you, Dad. I love you. Godspeed, Dad."[41]

NASA administrator Charles Bolden, a former shuttle commander, said Glenn "represented innovation and bravery, and with that infectious grin, he made us all feel good about ourselves." Looking back at Glenn's careers, Bolden said, "John Glenn always said 'yes.' Yes to his country's call in the US Marine Corps, yes to being the first American to orbit the Earth as one of the Mercury 7, yes to his state's nomination to serve in the Senate, and yes to the ongoing call of his nation to help to forge a path through a new millennium."[42] In reality, Glenn did more than say yes to challenges. In each case, he sought the chance to face difficult tasks and to serve his nation.

After returning from the service to Huntsville, Alabama, Ed Buckbee, an employee of NASA's public affairs office in the 1960s, told a reporter why he thought the White House had nixed another Glenn flight in the 1960s: "John Glenn was a national treasure." Buckbee described Glenn by saying, "He was one of those guys who had a vision for the future of NASA. . . . He gave you confidence when you were around him."[43] NBC News' reporter Jay Barbree said that Glenn's epitaph should be "Here lies a civilized man."[44]

Glenn was laid to rest in Arlington National Cemetery on a rainy April 6, 2017. His widow, Annie, had requested that the funeral be delayed until what would have been their seventy-fourth wedding anniversary. The drenching downpour added poignancy to this last farewell. A black, horse-drawn caisson carried Glenn's body to the burial site in a plain marine-issue casket he had requested himself.[45] The sound of the horses' hooves hitting the wet pavement rang out on an otherwise quiet day. Six white-gloved marine pallbearers carried the casket from the caisson to the gravesite, while the marine band played "The Marines' Hymn" at a slow, mournful tempo. The coffin was covered by an American flag covered in plastic to protect it from the rain.[46] A trumpeter played "Taps," marines fired three rifle rounds in his honor, and a minister led recitation of the Twenty-Third Psalm.[47] Then Robert B. Neller, commandant of the Marine Corps, knelt to hand the flag to ninety-seven-year-old Annie Glenn, who sat in a wheelchair and was wearing a red dress. He removed his hat and gloves and returned to give her a hug.[48] John Glenn held on to the "hero" label throughout the ceremony, and even years after death, it clings to him still. Three years after her husband, Annie Glenn died at 100 in May 2020.

Looking back at Glenn, colleagues recalled exactly the characteristics that his children had noted in their eulogy. After *Discovery*'s flight, the crew made appearances in many nations. Robinson said, "I saw him with kings and emperors and presidents, and I saw him with bus drivers and cooks and waiters, and he treated everyone the same." This quality, mentioned by many who knew him, was a central facet of John Glenn's being. His egalitarian streak combined with his courage to make him the unique individual he was.

Another set of consistently cited Glenn qualities were his curiosity and enthusiasm. "I think he's a guy that can take any experience and live it as fully as it could be lived," Robinson said. He recalled one of his favorite memories from Glenn's *Discovery* flight: Over South America cloaked by nighttime, lightning storms provided a dazzling light show. Glenn was staring out the window as he floated weightlessly. Robinson asked what he was doing, and without turning to answer, Glenn simply said, "I'm at church."[49] Glenn could crawl inside the beauty of the moment and savor it. Moments like this inspired astronaut, physician, and mountain climber Scott Parazynski to write, "When I grow up, I want to be like John. He's never stopped dreaming."[50] Beyond being a special man, Glenn was a terrific mentor, Robinson remembered. More than once after their shared flight, Robinson sought Glenn's advice on ideas that he wanted to implement. After thoroughly listening to the idea, he always said, "Let me think about that. I'll call you back tomorrow." On the next day, the call always came along with Glenn's well-considered thoughts.[51]

Even before Glenn's death, memory of his name was guaranteed.

Streets and schools across the country carry his name, as does the airport in Columbus. In addition to the John Glenn College of Public Affairs, Ohio State preserves his memory through the John Glenn Archives, which are part of the Ohio Congressional Archives. The John and Annie Glenn Museum, where Glenn's nine-room white-frame childhood home now stands, is in their hometown, New Concord.[52] The house, built by Glenn's father, was carried by flatbed truck to its new location on Main Street in 2001.[53] NASA operates the Glenn Research Center at Lewis Field in Cleveland, which does innovative research and design related to aeronautics and spaceflight.[54] For stargazers, there's the John Glenn Astronomy Park, which is part of Ohio's Hocking Hills State Park.[55]

One of NASA's commercial cargo suppliers to the International Space Station, Northrop Grumman Innovations Systems (formerly Orbital ATK), named a supply ship the SS *John Glenn*. The ship made its first and only visit to the orbiting outpost after blasting off from Cape Canaveral with seventy-seven hundred pounds of food, experiments, and other materials in 2017. It spent a while at the space station before departing and burning up in the atmosphere.[56] Blue Origin, a commercial spaceflight company, has given the former astronaut's name to a newly designed orbital launch vehicle. *New Glenn* is targeted to go into space for the first time in 2021.[57] On the island getaway spot of Turks and Caicos, a Splashdown Grand Turk exhibit celebrates Glenn's return to Earth near the island in 1962 and his brief stay on Grand Turk during debriefing following his Mercury flight.[58] A new incarnation of *The Right Stuff* as a TV series on Disney+ in 2020 again puts the spotlight on Glenn.

Glenn, the hero, had begun tying up his legacy long before his death. In 1962 "it seemed he had given Americans back their self-respect, and more than that—it seemed Americans dared again to hope," wrote Walter A. McDougall in *The Heavens and the Earth: A Political History of the Space Age*.[59] Tom Wolfe called him "the last true national hero America has ever had."[60] In the epilogue to his masterpiece, *The Right Stuff*, Wolfe wrote that the days "when an astronaut could parade up Broadway while traffic policemen wept in the intersections" had passed.[61] However, John Glenn's hero status, though somewhat diminished from 1962's level, was not missing in action when he made his second flight. Americans still came out to cheer his success.

Glenn had shown that a single feat does not define a hero. Instead, his achievements continued into a second act and a third. Did he make mistakes along the way? Of course, like all human beings, he struggled at times to gain his footing. Being a hero does not guarantee an ability to excel at every pursuit. It's not about being famous either. Many heroes live in anonymity, achieving incredible things that transform the lives of their families or their communities. For John Glenn, entering politics probably was more emotionally grueling than either of his spaceflights. For twenty-four years as a senator, he did his best, but he often found roadblocks in his path. He didn't have the politically savvy instincts to run a traditional presidential campaign. He wanted to ignore conventional wisdom and break new ground by not staging a typical race and not trying to spin his message. A naturally positive person, he wrestled with the issue of negative campaigning among Democrats competing for the nomination. Ultimately, the ingrained system of primaries and caucuses doomed his effort to triumph with a new kind of campaign. It was disappointing, and yet, failing to become president could not create a hole in a life that was so full. Similarly, while the Keating Five hubbub was demoralizing and embarrassing, its long-term effect was negligible.

When John Glenn flew in the shuttle *Discovery* at the end of his Senate service, he remained America's star astronaut; the often-dirty job of politics had not tainted that. Indeed, his eagerness to serve the public interest probably added a bit of incandescence to his story—so luminous that it was worthy of a place in mythology, although it was true. In some ways, it was Glenn's humanity, his fallibility that confirmed his status as a hero. His careful and determined labors to be a good man added polish to his daring deeds.

Glenn's gravestone puts his busy life into perspective. "If you see the headstone, you'll see that being a senator is the last thing that's listed," Rudy deLeon said. "You'll see that he was a combat pilot, he was an astronaut, and he was a senator."[62] In fact, Glenn's gravestone says:

John
Herschel
Glenn Jr.
Col.
US Marine Corps
WWII Korea
Jul 18, 1921
Dec 8, 2016
Fighter Pilot
Astronaut
US Senator

Glenn's achievements were great, but to many Americans, his name alone meant a great deal. Over more than half a century, he had signed thousands of autographs, whether he was walking within the US Capitol complex, dashing through an airport, moving from meeting to meeting at the Johnson Space Center, or simply enjoying a good meal. He always seemed willing to share that little bit of himself that could be summed up in his name. John Charles recalls one NASA session when Glenn had to dash out to fulfill another obligation. Before he left the room, he said, "I have to go. People are waiting for me. Does anybody else need an autograph?"[63] That was not ego talking. He recognized that his name alone had value to many Americans, not for its monetary worth or for its personal cachet, but for something totally different. The nine letters in his name said something about the boundless capacity of our nature—a greatness that often seems lost today. Glenn, like his autograph, was real, and yet amazing to behold.

DeLeon recalls being aboard the USS *John F. Kennedy* for a July Fourth ceremony in 2000, when Glenn "would see a line of twenty ship members wanting to come up, and he would always sign an autograph. And you'd see a line of twenty, and the next thing you saw, it would be a line of fifty. And he said he didn't mind at all. He said the taxpayers had funded the programs that were most important to him, like the Marine Corps and NASA, so he said it was simply an honor to have all these people walk up to him and say hello to him."[64] Butland believed "John handled fame better than any other famous person that I've ever encountered. . . . He never let it go to his head. He was always gracious and kind."[65]

Like the rest of us, Glenn wished for some things he never got— lunar dust on his boots, a seat in the Oval Office, and surely more. Nonetheless, he refused to feel cheated. Instead, he fully enjoyed the life he had been given and the people with whom he shared it. He offered unique gifts to the nation as a whole and especially to those gifted with the opportunity to know him personally. He remains an American hero, who now stands offstage, providing a scale against which succeeding generations of Americans can measure themselves.

In an article entitled "Post-Modern Blues," Marshall Fishwick wrote that heroes and legends "that served us for centuries have disappeared in cyberspace. We must work hard to find both mission and meaning in our tumultuous time—look for and welcome new views of who we are, and who we want to be. This may not be as hard as some think. The yearning for meaning—closely tied up in our myths, legends, and folklore—is always with us." He went on to predict that Americans would find heroes and heroines beyond "technocrats, sweepstake winners, professional athletes, and get-rich-quickers." He even suggested that real heroes would not depend on "spin doctors" as today's political leaders do.[66]

Sadly, the selflessness that John Glenn displayed as a marine, a forty-year-old astronaut, a senator, and a seventy-seven-year-old astronaut–guinea pig has become almost an alien concept in twenty-first-century America. Before John F. Kennedy told Americans, "Ask not what your country can do for you, ask what you can do for your country," Glenn already had embraced that ideal—and he never let it go. The me-me-me fixation that drives a significant number of current-day Americans casts a shadow over the nation's once-great ambitions.

Even curiosity seems to have been reduced to following celebrities. John Glenn was an explorer whether he was in space, in a Chinese opium den, or in a Senate hearing room. Too many Americans today care only about what's good for themselves and give little thought to their fellow Americans in need or to others like them in nations around the world.

In a time when fewer and fewer public figures attempt to be heroes or role models, a time when many prominent politicians don't even pretend to be guided by what's best for the public instead of what's best for their political party, John Glenn probably would be disappointed by the great nation he loved and served most of his life. He would hate the current political atmosphere. Polarization was his enemy; a united America, like the one he saw from space, was his goal. Those in power today follow a doctrine that offers no compromise, no unity, no heroes, and little hope for tomorrow.

What is John Glenn's legacy? "This world was a better place with John Glenn in it," said Robinson. "I do believe that everyone that he ever touched in his life became a better person—and ended up with a role model that they'll never lose."[67] That declaration can be advanced a step further by acknowledging that not just those who were fortunate enough to know him personally benefitted from his example. John Glenn showed Americans how to risk, how to sacrifice, how to win, how to lose, how to bounce back, how to grow old, and most importantly, how to live.

ACKNOWLEDGMENTS

I would like to thank the librarians at the John Glenn Archives, the John F. Kennedy Library, the National Archives in Fort Worth, the Paley Center for Media in New York, and NASA headquarters. I also want to express my gratitude to Jim Banke, my technical editor, and the researchers who helped me with long-distance research at other libraries, especially Denise Gamino in Austin. In addition, I am grateful to my agent Roger Williams and my editors at Chicago Review Press—Jerome Pohlen, Michelle Williams, and Benjamin Krapohl. Finally, I want to thank my proofreader, Lou Oschmann.

NOTES

LIBRARY ABBREVIATIONS

HRC—Historical Reference Collection, NASA Headquarters, Washington, DC
JCL—Jimmy Carter Library, Atlanta, GA
JFKL—John F. Kennedy Library, Boston, MA
JGA—John Glenn Archives, Ohio Congressional Archives, Ohio State University, Columbus, OH
LBJL—Lyndon B. Johnson Library, Austin, TX
NA/II—National Archives II, College Park, MD
NA/FW—National Archives, Fort Worth, TX
PCFM—Paley Center for Media, New York, NY
WJCL—William Jefferson Clinton Library, Little Rock, AR

PROLOGUE

1 Michael Neufeld, interview by author, June 13, 2019.
2 Tom Wolfe, *The Right Stuff* (New York: Bantam Books, 1979), 277.
3 Arthur Krock, "Another Talent Revealed Today by Colonel Glenn," *New York Times*, February 27, 1962.
4 "Colonel Wonderful," *Time*, March 9, 1962, 22.
5 Kennedy first set the goal in a speech to a joint session of Congress in May 1961, but the Rice speech is best remembered.
6 Jade Boyd, "JFK's 1962 Moon Speech Still Appeals 50 Years Later," *Rice University News and Media*, August 30, 2012, http://news.rice.edu/2012/08/30/jfks-1962-moon-speech -still-appeals-50-years-later.
7 Of course, there is one major difference between western expansion and space exploration: An indigenous population had to be moved out of the way to make room for frontier settlers of mostly European descent. In space, at least so far, the United States is free to advance without denying rights to another population.
8 Asif A. Saddiqi, "Spaceflight in the American Imagination," in *Remembering the Space Age: Proceedings of the 50th Anniversary Conference*, ed. Steven J. Dick (Washington, DC: National Aeronautics and Space Administration, 2008), 24.
9 Neufeld, interview by author.

10 Wolfe, *Right Stuff*, 97.

11 George Dixon, "Washington Scene . . . Drama Needs a Hero," 1962, King Features Syndicate, Non-Senate Papers Subgroup, Box 24, Folder 5, JGA.

12 Saul Pett, "Expert Probes U.S. Tumult over Its Hero, John Glenn," Non-Senate Papers Subgroup, Box 3, Folder 32, JGA.

13 Father Quinn, "What Is a Hero?" Non-Senate Papers Subgroup, Box 23, Folder 9, JGA.

14 John Noble Wilford, "50 Years Later, Celebrating John Glenn's Feat," *New York Times*, February 14, 2012.

15 Mark E. Byrnes, *Politics and Space: Image Making by NASA* (Westport, CT: Praeger, 1994), 54.

16 Dale Butland, interview by author, July 17, 2019.

17 Committee on Science and Astronautics, US House of Representatives, Eighty-Sixth Congress, Second Session, *The Practical Values of Space Exploration* (Washington, DC: Government Printing Office, 1960), 52.

18 John B. Charles, interview by author, January 14, 2019.

19 Steve Robinson, interview by author, February 13, 2019.

20 Robinson, interview by author.

21 Charles, interview by author.

22 Byrnes, *Politics and Space*, 55.

23 Roger D. Launius, "American Spaceflight History's Master Narrative and the Meaning of Memory," in Dick, *Remembering the Space Age*, 360.

24 Roger D. Launius, "Heroes in a Vacuum: The Apollo Astronaut as Cultural Icon," *Florida Historical Quarterly*, Fall 2008, 193.

25 Pett, "Expert Probes U.S. Tumult." Walter Mitty was a fictional character created by James Thurber. Mitty first appeared in a short story by Thurber published in the *New Yorker* and later featured in two films, both entitled *The Secret Life of Walter Mitty*. An ordinary man, Mitty has fantasies of doing extraordinary things.

26 John Dille, introduction to *We Seven*, by M. Scott Carpenter, L. Gordon Cooper Jr., John H. Glenn Jr., Virgil L. Grissom, Walter M. Schirra Jr., Alan B. Shepard Jr., and Donald K. Slayton (New York: Simon & Schuster, 1962), 13.

27 Wally Schirra with Richard N. Billings, *Schirra's Space* (Annapolis: Naval Institute Press, 1988), 71.

28 Scott Carpenter with Kris Stoever, *For Spacious Skies: The Uncommon Journey of a Mercury Astronaut* (New York: New American Library, 2004), 209.

29 Butland, interview by author; Ron Grimes, interview by author, July 12, 2019.

30 Kathy Connolly, interview by author, March 7, 2019.

31 John Haseley, interview by author, July 16, 2019.

32 Grimes, interview by author.

33 Daniel S. Goldin, "Toast to John Glenn," September 14, 1998, Historical Reference Collection Online Database, program office, NASA Historian, Record No. 32186, HRC.

34 Robinson, interview by author.

35 "36 Years Later, We Still Hunger for Heroes," *USA Today*, October 29, 1998.

36 Ben Wattenberg, "National Heroism Void?" *Washington Times*, August 12, 1999.

37 William Kristol, quoted in Wattenberg, "National Heroism Void?"

38 David McCullough, quoted in Peter H. Gibbon, *A Call to Heroism: Renewing America's Vision of Greatness* (New York: Grove, 2007), 114–115.

39 Adam Gopnik, quoted in Gibbon, *Call to Heroism*, 93.

40 Barbara Tuchman, quoted in Wattenberg, "National Heroism Void?"

41 Thomas Jefferson, quoted in Gibbon, *Call to Heroism*, 117.

42 McCain, who was held as a North Vietnamese prisoner of war for more than six years, is a hero to some, but President Donald Trump has belittled his record, even after McCain's death.

43 "Veteran Homeless Facts," Green Doors, accessed February 15, 2019, https://greendoors.org/facts/veteran-homelessness.php.

44 Edmund Morris, *The Rise of Theodore Roosevelt* (New York: Random House, 1979), 684.

45 Jason Pramuk, "Trump: President Barack Obama Was Born in the United States. Period," CNBC, September 26, 2016, www.cnbc.com/2016/09/16/trump-president-obama-was -born-in-the-united-states-period.html.

46 Sara Terry, "Why Heroes Are Hard to Come By," *Christian Science Monitor*, November 9, 1998, www.csmonitor.com/1998/1109/110998.us.us.2.html.

47 James Reston, "Cape Canaveral: Is the Moon Really Worth John Glenn?" *New York Times*, February 25, 1962.

48 "Public Trust in Government, 1958–2019," Pew Research Center, April 11, 2019, www.people-press.org/2019/04/11/public-trust-in-government-1958-2019.

49 William R. Newcott, "John Glenn: Man with a Mission," *National Geographic*, June 1999, 66.

CHAPTER 1: SMALL-TOWN BOY

1 John H. Glenn Jr., oral history interview 1, by Brien R. Williams, October 25, 1996, 68, John H. Glenn Jr. Oral History Project, John Glenn Oral History Project, JGA.

2 William R. Shelton, "A Year Book Special Report," *World Book Year Book 1963* (Chicago: Field Enterprises, 1963), 163.

3 John H. Glenn Jr., untitled and undated class essay, Non-Senate Papers Subgroup, Box 1, Folder 17, JGA.

4 Glenn, oral history interview 1, Williams, 5.

5 Glenn, oral history interview 1, Williams, 17.

6 Glenn, oral history interview 1, Williams, 47.

7 Glenn, oral history interview 1, Williams, 19.

8 Glenn, oral history interview 1, Williams, 4.

9 Glenn, oral history interview 1, Williams, 37.

10 John H. Glenn Jr., oral history interview 2, by Brien R. Williams, October 28, 1996, 4, John H. Glenn Jr. Oral History Project, JGA.

11 See John Glenn with Nick Taylor, *John Glenn: A Memoir* (New York: Bantam Books, 1999), 12. As an adolescent, she "struggled with school and with other aspects of her life," in Glenn's words. At that point, she disappears from Glenn's memoir (see Glenn and Taylor, 83). In a 1996 oral history, when she would have been about seventy, Glenn reported that she was in a Tennessee nursing home. See Glenn, oral history interview 1, Williams, 46.

12 Glenn, oral history interview 2, Williams, 2–3.

13 Glenn, oral history interview 2, Williams, 29–30.

14 John H. Glenn Jr., oral history interview 4, by Brien R. Williams, December 12, 1996, 10–15, John H. Glenn Jr. Oral History Project, JGA.

15 So soon after the Great Influenza Pandemic of 1918–1919, which killed around 650,000 Americans, public officials were especially conscious of the dangers of diseases spreading.

16 John H. Glenn Jr., oral history interview, by Sheree Scarborough, August 25, 1997, 1, National Aeronautics and Space Administration Oral History Project, https:// historycollection.jsc.nasa.gov/JSCHistoryPortal/history/oral_histories/GlennJH /GlennJH_8-25-97.htm.

17 In an age before airline manufacturers merged into giant corporations, the two-seater was manufactured by Ohio-based Waco Aircraft Company.

18 Glenn with Taylor, *John Glenn*, 14–15.

19 "Grade School News," *Barnesville (OH) Enterprise*, May 26, 1932, Non-Senate Papers Subgroup, Box 1, Folder 19, JGA.

20 Glenn, oral history interview 1, Williams, 10–11.

21 Glenn, oral history interview 1, Williams, 14–15.

22 Rudy deLeon, interview by author, June 10, 2019.

23 Glenn oral history interview 1, Williams, 27-28.

24 Glenn, oral history interview 1, Williams, 31–32.

25 Jeffrey Kluger, "Liftoff: The World Watched as John Glenn Launched in 1962 and Time Presented Unparalleled Cover on the Story," in "John Glenn: A Hero's Life, 1921–2016," ed. Nancy Gibbs, commemorative edition, *Time*, 2016, 24.

26 C. Edwin Houk, "A Spaceman Who Lives by Facts and Faith," *Light*, October 7, 1962, 1–2.

27 Houk, "Spaceman Who Lives," 1–2.

28 Glenn, oral history interview 2, Williams, 7–8.

29 Glenn, oral history interview 1, Williams, 18.

30 Glenn, oral history interview 1, Williams, 78–79.

31 Annie Glenn, oral history interview 2, by Brien R. Williams, November 7, 1997, 5, John H. Glenn Jr. Oral History Project, JGA.

32 Glenn, oral history interview 1, Williams, 68–69.

33 Annie Glenn, oral history interview 2, Williams, 8.

34 Glenn, oral history interview 2, Williams, 10–12.

35 Glenn, oral history interview 1, Williams, 35–36, 42–44.

36 Glenn, oral history interview 1, Williams, 37–38.

37 Glenn, oral history interview 1, Williams, 66–67.

38 Glenn, oral history interview 1, Williams, 75.

39 Glenn with Taylor, *John Glenn*, 55–56.

40 Houk, "Spaceman Who Lives," 3.

41 Glenn, oral history interview 1, Williams, 60–62.

42 Shelton, "Year Book Special Report," 163.

43 John H. Glenn Jr., "The Choice of a Mate," term paper, January 20, 1942, 5, Non-Senate Papers Subgroup, Box 2, Folder 3, JGA.

44 DeLeon, interview by author.

CHAPTER 2: OFF TO WAR

1 John H. Glenn Jr., oral history interview 8, by Brien R. Williams, April 7, 1997, 7, John H. Glenn Jr. Oral History Project, JGA.

2 John H. Glenn Jr., oral history interview 3, by Brien R. Williams, December 6, 1996, 65–71, John H. Glenn Jr. Oral History Project, JGA.

3 Glenn, oral history interview 1, Williams, 76.

4 Glenn, oral history interview 3, Williams, 73.

5 Glenn, oral history interview 3, Williams, 74–75.

6 Glenn, oral history interview 4, Williams, 16.

7 Ensign R. S. Carlson, "Officer Aptitude Report for Student Officers or Cadets," August 22, 1942, Non-Senate Papers Subgroup, Box 13, Folder 20, JGA.

8 Annie Glenn, oral history interview 2, Williams, 9.

9 Glenn, oral history interview 3, Williams, 82.

10 Glenn, oral history interview 3, Williams, 67.

11 Glenn, oral history interview 4, Williams, 25–26.

12 John H. Glenn Jr., diary, March 10, 1944 through undetermined date, 7, Non-Senate Papers Subgroup, Box 13, Folder 33.

13 Glenn, oral history interview 4, Williams, 30.

14 Glenn, oral history interview 4, Williams, 28.

15 Glenn, oral history interview 4, Williams, 36–38.

16 Glenn, oral history interview 4, Williams, 31–32.

17 Glenn, diary, 9.

18 Glenn, diary, 14.

19 Glenn, diary, 13.

20 Glenn and Taylor, *John Glenn*, 104.

21 Glenn, oral history interview, Scarborough, 1.

22 Glenn, diary, 10–11.

23 Glenn, diary, 16.

24 Glenn with Taylor, *John Glenn*, 108.

25 Glenn, diary, 21.

26 John H. Glenn Jr., oral history interview 5, by Brien R. Williams, January 16, 1997, 3–4, John H. Glenn Jr. Oral History Project, JGA.

27 Glenn, diary, 22–25.

28 John H. Glenn Jr., oral history interview 6, by Brien R. Williams, February 7, 1997, 10, John H. Glenn Jr. Oral History Project, JGA.

29 "Battle of Midway," History.com, original: October 29, 2009, updated: December 17, 2019, www.history.com/topics/world-war-ii/battle-of-midway.

30 Glenn, diary, 30–31.

31 Glenn, diary, 27.

32 Glenn, diary, 34–36.

33 Glenn, oral history interview 5, Williams, 5–6.

34 Glenn, oral history interview 5, Williams, 9–10.

35 Glenn, diary, 37.

36 John H. Glenn Jr., oral history interview 7, by Brien R. Williams, February 10, 1997, 8, John H. Glenn Jr. Oral History Project, JGA.

37 Glenn, oral history interview 5, Williams, 7.

38 Glenn, oral history interview 5, Williams, 8.

39 Glenn, diary, 41–43.

40 Glenn, oral history interview 5, Williams, 13.

41 Glenn, oral history interview 6, Williams, 15.

42 Glenn, diary, 50–51.

43 Glenn, oral history interview 5, Williams, 16–17.

44 Glenn with Taylor, *John Glenn*, 123.

45 Wireless reports, "Japan's Phosphates Menaced by Allies," *New York Times*, February 27, 1944.

46 Associated Press, "U.S. Planes Strike Pagan, Rota, Nauru," *New York Times*, August 21, 1944.

47 Glenn, oral history interview 5, Williams, 35–36.

48 Kluger, "Liftoff," 26.

49 Glenn, oral history interview 5, Williams, 31.

50 Glenn, oral history interview 5, Williams, 19.

51 Glenn, oral history interview 5, Williams, 18.

52 Bill Flanagan, "A Little-Known Story from the Life of John Glenn," CBS News, December 11, 2016, www.cbsnews.com/news/a-little-known-story-from-the-life-of-John-Glenn.

53 Dan Hampton, *The Flight: Charles Lindbergh's Daring and Immortal 1927 Transatlantic Flight* (New York: HarperCollins, 2017), 266–267.

54 "Napalm," United States History (website), accessed July 8, 2017, www.u-s-history.com/pages/h1859.html.

55 Glenn with Taylor, *John Glenn*, 135.

56 John Glenn, "Parts Taken from John's Letters," 1, Non-Senate Papers Subgroup, Box 14, Folder 3, JGA.

57 Glenn, oral history interview 6, Williams, 41–42.

58 Glenn with Taylor, *John Glenn*, 136.

59 Glenn, oral history interview 6, Williams, 23–24.

60 Glenn, "Parts Taken from John's Letters," 1.

61 Glenn, oral history interview 6, Williams, 13.

62 Glenn, oral history interview 5, Williams, 32.

63 "Fourth Marine Aircraft Wing," GlobalSecurity.org, accessed May 24, 2019, www.globalsecurity.org/military/agency/usmc/4maw.htm.

64 US Marine Corps, "War Diary," covering January 1–31, 1944, undated entry, 4, Non-Senate Papers Subgroup, Box 14, Folder 8, JGA.

65 "Recommendation for Award-Strike/Flight System," Non-Senate Papers Subgroup, Box 14, Folder 24, JGA.

66 "Marine Fighting Squadron One Fifty-Five," Non-Senate Papers Subgroup, Box 14, Folder 8, JGA.

67 Glenn with Taylor, *John Glenn*, 137.

68 Glenn, oral history interview 5, Williams, 43.

69 Glenn, oral history interview 7, Williams, 1.

70 Senator John Glenn, "We Will," December 7, 1991, Senate Papers, Box 105.2, Folder 8, JGA.

71 John H. Glenn Jr., oral history interview 10, by Brien R. Williams, May 12, 1997, 7, John H. Glenn Jr. Oral History Project, JGA.

72 Glenn, oral history interview 10, Williams, 18.

73 Glenn, oral history interview 10, Williams, 14–15.

74 Glenn, oral history interview 10, Williams, 12.

75 Glenn, oral history interview 10, Williams, 14.

76 Glenn, oral history interview 10, Williams, 23.

77 Glenn, oral history interview 10, Williams, 7.

78 Glenn, oral history interview 10, Williams, 25.

79 Glenn, oral history interview 10, Williams, 28.

80 Glenn, oral history interview 10, Williams, 28–31.

81 John H. Glenn Jr., oral history interview 11, by Brien R. Williams, May 12, 1997, 4, John H. Glenn Jr. Oral History Project, JGA.

82 Glenn, oral history interview 11, Williams, 13.

83 Glenn, oral history interview 11, Williams, 15.

84 Captain John H. Glenn Jr., report to chief of Naval Air Advanced Training, September 8, 1950, Non-Senate Papers Subgroup, Box 14, Folder 61, JGA.

85 Glenn, oral history interview 11, Williams, 17–18.

86 Glenn, oral history interview 11, Williams, 20.

87 Glenn, oral history interview 11, Williams, 20–22.

88 Glenn, oral history interview 11, Williams, 42.

89 Glenn, oral history interview 11, Williams, 35.

90 "Korean War," History.com, original: November 9, 2009, updated: January 21, 2020, www.history.com/topics/korea/korean-war.

91 C. N. Trueman, "The United Nations and the Korean War," History Learning Site, May 26, 2015, www.historylearningsite.co.uk/modern-world-history-1918-to-1980/the-united-nations/the-united-nations-and-the-korean-war.

92 John H. Glenn Jr., oral history interview 14, by Brien R. Williams, March 13, 1998, 2–3, John H. Glenn Jr. Oral History Project, JGA.

CHAPTER 3: OL' MAGNET TAIL

1 Glenn, oral history interview 2, Williams, 30.

2 Glenn, oral history interview 14, Williams, 1.

3 "Ted Williams Stats," Baseball Almanac, last updated August 30, 2019, www.baseball-almanac.com/players/player.php?p=willite01.

4 Michael Seidel, *Ted Williams: A Baseball Life* (South Orange, NJ: Summer Games Books, 2015), 88.

5 Seidel, *Ted Williams*, 203.

6 Glenn, oral history interview 14, Willliams, 7.

7 Glenn, oral history interview 14, Willliams, 8.

8 Glenn, oral history interview 14, Willliams, 10.

9 Glenn, oral history interview 14, Willliams, 22.

10 Newcott, "John Glenn," 64.

11 Glenn, oral history interview 14, Williams, 24–25.

12 Glenn, oral history interview 14, Williams, 33.

13 John Glenn, special release to Naval Air News, Non-Senate Papers Subgroup, Box 16.1, Folder 6, JGA.

14 John H. Glenn Jr., oral history interview 15, by Brien R. Williams, March 23, 1998, 1–2, John H. Glenn Jr. Oral History Project, JGA.

15 Kluger, "Liftoff," 26.

16 Lily Koppel, *The Astronaut Wives Club: A True Story* (New York: Grand Central, 2013), 77.

17 Jeffrey Kluger, "An American Icon," in Gibbs, "John Glenn," 16.

18 Peter B. Mersky, *U.S. Marine Corps Aviation Since 1912* (Annapolis: Naval Institute Press, 2009), 150.

19 Glenn, oral history interview 4, Williams, 49–50.

20 Glenn, oral history interview 14, Williams, 10.

21 Glenn, oral history interview 4, Williams, 57.

22 Glenn with Taylor, *John Glenn*, 172.

23 Eloise Engle, "What a Fireball!" *Listen*, July–August 1962, 18.

24 Leigh Montville, *Ted Williams: The Biography of an American Hero* (New York: Broadway Books, 2005), 168.

25 John Glenn, "Information Please," *Marine Corps Gazette*, Non-Senate Papers Subgroup, Box 16.1, Folder 2, JGA.

26 John Glenn, "Before the Run," Non-Senate Papers Subgroup, Box 16.1, Folder 2, JGA.

27 John Glenn, "An Interview," Non-Senate Papers Subgroup, Box 16.1, Folder 4, JGA.

28 J. A. Pounds III, letter to John Glenn, March 26, 1951, Non-Senate Papers Subgroup, Box 15, Folder 4, JGA.

29 Glenn, oral history interview 14, Williams, 13.

30 Glenn, oral history interview 15, Williams, 15–16.

31 Glenn, oral history interview 15, Williams, 6.

32 Glenn, oral history interview 15, Williams, 8.

33 Glenn, oral history interview 15, Williams, 11–12.

34 Brian R. Swopes, "23 July 1953," *This Day in Aviation* (blog), July 23, 2019, www.thisdayinaviation.com/tag/mig-mad-marine.

35 Glenn, oral history interview 15, Williams, 22–30.

36 Glenn with Taylor, *John Glenn*, 193.

37 Major John H. Glenn Jr., memo to commandant of the Marine Corps, November 3, 1953, appendix 1, page 2–3, Non-Senate Papers Subgroup, Box 16.1, Folder 1, JGA.

38 John Glenn, unidentified report on air force experiences, undated, Non-Senate Papers Subgroup, Box 16.1, Folder 5, JGA.

39 Glenn, oral history interview, Scarborough, 4–5.

40 "Korean War," History.com.

41 John H. Glenn Jr., oral history interview 16, by Brien R. Williams, March 27, 1998, 1, John H. Glenn Jr. Oral History Project, JGA.

42 Glenn, oral history interview 16, Williams, 11.

43 Glenn, oral history interview 16, Williams, 11.

44 Glenn, oral history interview 16, Williams, 13.

45 Glenn, oral history interview 16, Williams, 13–14.

46 Glenn, oral history interview 16, Williams, 15.

47 Glenn, oral history interview 16, Williams, 16–17.

48 Glenn, oral history interview 16, Williams, 17–19.

49 Glenn, oral history interview 16, Williams, 20–22.

50 "F7U Cutlass," National Naval Aviation Museum, accessed November 10, 2018, www.navalaviationmuseum.org/aircraft/f7u-cutlass.

51 John H. Glenn Jr., oral history interview 17, by Jeffrey W. Thomas, March 7, 2008, 3, John Glenn Oral History Project, JGA.

52 Glenn, oral history interview 17, Thomas, 4.

53 Chris Kraft, *Flight: My Life in Mission Control* (New York: Penguin Books, 2001), 53–56.

54 Glenn, oral history interview 17, Thomas, 6.
55 Glenn, oral history interview 17, Thomas, 8.
56 Glenn, oral history interview 17, Thomas, 11–12.
57 Glenn, oral history interview 17, Thomas, 12–17.
58 Glenn, oral history interview 17, Thomas, 17–18.
59 Kluger, "Liftoff," 27.
60 Harrison E. Salisbury, "Jet Flier Crosses U.S. in Record 3 Hours 23 Minutes," *New York Times*, July 17, 1957.
61 A. Schellhammer, "F8U-1 Flight," memorandum of Chance Vought aircraft, June 26, 1957, 1, Non-Senate Papers Subgroup, Box 18, Folder 20, JGA.
62 Salisbury, "Jet Flier Crosses U.S."
63 Glenn, oral history interview 17, Thomas, 19–20.
64 "A Real Speedy Aviator: John Hershall Glenn Jr.," *New York Times*, July 17, 1957.
65 Glenn, oral history interview 17, Thomas, 21.
66 Secretary of the Navy Thomas S. Gates Jr., "Distinguished Flying Cross Citation," July 18, 1957, Non-Senate Papers Subgroup, Box 13, Folder 2, JGA.
67 Ted Williams, telegram to John Glenn, undated, Non-Senate Papers Subgroup, Box 18, Folder 27, JGA.
68 "Real Speedy Aviator," *New York Times*.
69 John Innes, "Fast Pass," *Vought Vanguard*, July 24, 1957, 1, Non-Senate Papers Subgroup, Box 18, Folder 17, JGA.
70 Glenn, oral history interview 17, Thomas, 22.
71 *Name That Tune*, Columbia Broadcasting System, September 24, 1957, PCFM.
72 "Eddie & the Major," *Name That Tune* transcript, Non-Senate Papers Subgroup, Box 18, Folder 16, JGA.
73 Mildred Willard, letter to John Glenn, September 18, 1957, Non-Senate Papers Subgroup, Box 18, Folder 21, JGA.

CHAPTER 4: RACING TOWARD SPACE

1 Over the years of the space race, the Soviet Union sent many dogs into space. Laika was the only one launched with no opportunity for survival. The flight was planned quickly to coincide with the fortieth anniversary of the Bolshevik revolution, and there was no way to provide a safe recovery of the dog, a husky-spitz mix. See Alice George, "The Sad, Sad Story of Laika, the Space Dog, and Her One-Way Trip into Orbit," *Smithsonian Magazine*, April 11, 2018, www.smithsonianmag.com/smithsonian-institution/sad-story-laika-space-dog-and-her-one-way-trip-orbit-1-180968728.
2 "Vanguard TV3," National Aeronautics and Space Administration, accessed May 28, 2019, https://nssdc.gsfc.nasa.gov/nmc/spacecraft/display.action?id=VAGT3.
3 "Explorer 1 Overview," in "Explorer 1 and Early Satellites," National Aeronautics and Space Administration, accessed May 28, 2019, www.nasa.gov/mission_pages/explorer/explorer-overview.html.
4 Glenn with Taylor, *John Glenn*, 240–241.
5 John H. Glenn Jr., oral history interview 18, by Jeffrey W. Thomas, April 21, 2008, 15, John H. Glenn Jr. Oral History Project, JGA.
6 George M. Low, "Pilot Selection for Project Mercury," memorandum for administrator, April 23, 1959, 1, Record 6816, Astronauts Folder, HRC.
7 Low, "Pilot Selection," 1.
8 "Project Mercury Overview—Astronaut Selection," National Aeronautics and Space Administration, November 30, 2006, www.nasa.gov/mission_pages/mercury/missions/astronaut.html.
9 Glenn, oral history interview, Scarborough, 6–7.
10 Wolfe, *Right Stuff*, 72–73.
11 Johannes Kepler, *Somnium* (Middletown, Del.: Amazon, 2019), 14.

12 Jeffrey Kluger, *Apollo 8: The Thrilling Story of the First Mission to the Moon* (New York: Henry Holt, 2017), 48.

13 Jim Lovell and Jeffrey Kluger, *Apollo 13* (Boston: Houghton Mifflin, 1994), 180–183.

14 Glenn, oral history interview, Scarborough, 8.

15 Neal Thompson, *Light This Candle: The Life and Times of Alan Shepard* (New York: Three Rivers, 2005), 195.

16 Glenn, oral history interview 18, Thomas, 9–10.

17 "Sample Questions from Project Mercury Tests," news release, National Aeronautics and Space Administration, May 12, 1959, 1, Record 6818, 7 Astronauts Folder, HRC.

18 Wolfe, *Right Stuff*, 76–77.

19 Koppel, *Astronaut Wives Club*, 19.

20 Sheldon J. Korchin and George E. Ruff, "Personality of Mercury Astronauts," in *The Threat of Impending Disaster*, ed. George H. Grosser, Henry Wechsler, and Milton Greenblatt (Cambridge: Massachusetts Institute of Technology, 1964), 200–203.

21 Alan Shepard and Deke Slayton with Jay Barbee, *Moon Shot* (New York: Open Road, 2011), 51.

22 "Space Voyagers Rarin' to Orbit," *Life*, April 26, 1959, 22.

23 Press conference: Mercury Astronaut Team, National Aeronautics and Space Administration, April 9, 1959, 6, National Aeronautics and Space Administration, Records of History Office, Chronological Files, Box 7, February–April 16, 1959 folder, NA/FW.

24 Press conference: Mercury Astronaut Team, 7.

25 Matthew H. Hersch, *Inventing the American Astronaut* (New York: Palgrave Macmillan, 2012), 27.

26 Kluger, *Apollo 8*, 45.

27 Press conference: Mercury Astronaut Team, 16–19.

28 Press conference: Mercury Astronaut Team, 22.

29 Walter M. Schirra Jr., oral history interview, by Roy Neal, December 1, 1998, 20–21, Johnson Space Center Oral History Project.

30 John A. Powers, memorandum to astronauts, April 24, 1959, 1, National Aeronautics and Space Administration Records, Lyndon B. Johnson Space Center, office of the director, Box 18, Colonel Powers—Public information officer folder, NA/FW.

31 "*I've Got a Secret* Trivia," IMDb, accessed March 16, 2019, www.imdb.com/title /tt0044270/trivia.

32 "Astronaut Program Outlined," press release, National Aeronautics and Space Administration, May 12, 1959, Record 6018, 7 Astronauts Folder, NASA Headquarters, HRC.

33 Kluger, "Liftoff," 27.

34 Ray E. Boomhower, *Gus Grissom: The Lost Astronaut* (Indianapolis: Indiana Historical Society Press, 2004), 127.

35 Transcript of Martin Bush's tape-recorded interview with four of the seven astronauts, March 25, 1960, 1–2, National Aeronautics and Space Administration, Records of History Office, Chronological Files, Box 11, March 18–31, 1960, NA/FW.

36 Transcript of Bush interview, 5.

37 Transcript of Bush interview, 10–11.

38 George C. Guthrie, memorandum for project director, May 28, 1959, 7, National Aeronautics and Space Administration, Records of History Office, Chronological Files, Box 3 of 4, April 17– May 31 folder, NA/FW.

39 Thompson, 217.

40 Glenn, oral history interview 18 by Jeffrey Thomas, John H. Glenn Jr. Oral History Project, April 21, 2008, 18, JGA.

41 Wolfe, 110–111.

42 "Summary of the Mercury-Johnsville centrifuge program of August 1959," memorandum for chief, November 16, 1959, 3, National Aeronautics and Space Administration, Records of the Office of History, Chronological Files, Box 8, November 6–16 folder, National Archives/Fort Worth.

43 Glenn, oral history interview 18 by Jeffrey Thomas, 41.

44 Robert B. Voas, "Project Mercury: The Astronaut Training Program," presentation at the Symposium on Psychophysiological Aspects of Space Flight, May 26–27, 1960, Non-Senate Papers Subgroup, Box 60, Folder 19, JGA.

45 Hersch, 23.

46 M. Scott Carpenter, oral history interview 2 by Roy Neal, Johnson Space Center Oral History Project, January 27, 1999, 19.

47 Kluger, "An American Icon," 27.

48 Wolfe, 111.

49 Thompson, *Light This Candle*, 267–269.

50 Glenn with Taylor, *John Glenn*, 293–295.

51 Wolfe, *Right Stuff*, 140.

52 "Astronaut Desert Survival Training Program," memorandum for associate director, October 7, 1960, 1, Non-Senate Papers Subgroup, Box 69, Folder 35, JGA.

53 Donald K. Slayton, "Astronaut Comments on Desert Survival Training," 3, Non-Senate Papers Subgroup, Box 69, Folder 35, JGA.

54 Glenn with Taylor, *John Glenn*, 299.

55 John H. Glenn, "Astronaut Comments on Desert Survival Training," 1–2, Non-Senate Papers Subgroup, Box 69, Folder 35, JGA.

56 Wolfe, *Right Stuff*, 172.

57 Alan B. Shepard Jr., oral history interview, by Roy Neal, February 20, 1998, 6, Johnson Space Center Oral History Project.

58 Thompson, *Light This Candle*, 273.

59 Wolfe, *Right Stuff*, 180–181.

60 Boomhower, *Gus Grissom*, 156.

61 Looking back, Glenn said more than forty years later, "The way it all worked out with the Russians and the scheduling, and the way the flight pattern went, why it was a good thing that I was on the third flight. I have no complaints today." See Glenn, oral history interview 18, Thomas, 34.

62 George J. Feldman and Charles S. Sheldon II, "Interim Report on Space Policy," April 24, 1961, 30, Lyndon B. Johnson Papers, Vice Presidential Security File, Box 17, Space Program— Secret Folder, LBJL.

63 Feldman and Sheldon, "Interim Report on Space Policy," 29.

64 This premise was confirmed most clearly in Apollo 8's voyage to orbit the moon, seven years after this report had been written. The flight confirmed American ambitions to reach out to the moon, and the mission's astronauts used their human eyes to recognize the beauty of Earth as it rose over a barren lunar landscape and to capture their gorgeous, delicate planet in photographs that would fundamentally change the way millions of humans envisioned their home. Apollo 8 images, captured by the first humans to leave Earth in the rearview mirror and voyage to another body in space, enhanced protective feelings toward a fragile home world. On the front page of the *New York Times*, poet Archibald MacLeish wrote that images recorded by Apollo 8 astronauts could reconfigure human thought: "To see the Earth as it truly is, small and blue and beautiful in the eternal silence where it floats, is to see ourselves as riders on the Earth together, brothers on that bright loveliness in the eternal cold—brothers who know now they are truly brothers." The most special photo, "Earthrise" by astronaut Bill Anders, became a centerpiece of the environmental movement. See Archibald MacLeish, "A Reflection: Riders on Earth Together, Brothers in Eternal Cold," *New York Times*, December 25, 1968.

65 Edward R. Murrow, memorandum to McGeorge Bundy, April 3, 1961, JFKL, www.jfklibrary.org/Asset-Viewer/Archives/JFKNSF-307-004.aspx.

66 Thompson, *Light This Candle*, 277–278.

67 Thompson, 285–286.

68 Many Americans believed the 7 stood for the seven astronauts, so all future Mercury

flights included the number in their names. In truth, it was *Freedom 7* because it was the seventh capsule to roll off the line. See M. Scott Carpenter, oral history interview 2, by Roy Neal, January 27, 1999, 9, Johnson Space Center Oral History Project.

69 Thompson, *Light This Candle*, 289–290.

70 Shepard, oral history interview, Neal, 8.

71 "Shepard Medal Presentation," YouTube video, posted by "Mark Gray," May 5, 2013, https://youtu.be/4Qd-t3vFsxc.

72 Shepard and Slayton with Barbee, *Moon Shot*, 120.

73 Bureau of the Budget, "Substantive Objectives of U.S. Space Programs," May 5, 1961, 1, JFKL, www.jfklibrary.org/asset-viewer/archives/JFKNSF/307/JFKNSF-307-004.

74 James E. Webb and Robert McNamara, "Recommendations for our National Space Program: Changes, Policies, Goals," May 8, 1961, 7, JFKL, www.jfklibrary.org/asset-viewer/archives/JFKNSF/307/JFKNSF-307-004.

75 John F. Kennedy, "Excerpt from the 'Special Message to the Congress on Urgent National Needs,'" National Aeronautics and Space Administration, May 25, 1961, www.nasa.gov/vision/space/features/jfk_speech_text.html.

76 It was recovered in an operation funded by the Discovery Channel in 1999. See "Liberty Bell 7 Capsule Raised from Ocean Floor," CNN, July 20, 1999, www.cnn.com/TECH/space/9907/20/grissom.capsule.01.

77 Space Science Board of National Academy of Sciences, "Man's Role in the National Space Program," August 1961, 2, National Aeronautics and Space Administration, Records of the History Office, Chronological Files, Box 19, August 1961 folder, NA/FW.

78 Schirra with Billings, *Schirra's Space*, 72.

79 Donald K. Slayton with Michael Cassutt, *Deke!* (New York: Forge, 1994), 101.

CHAPTER 5: AROUND THE WORLD IN 89 MINUTES

1 Walter Cunningham, *The All-American Boys: An Insider's Look at the U.S. Space Program* (New York: ipicturebooks, 2009), 336.

2 John Glenn, "'If You're Shook Up, You Shouldn't Be There,'" *Life*, March 9, 1962, 26.

3 Boomhower, *Gus Grissom*, 222.

4 Loyd S. Swenson Jr., James M. Grimwood, and Charles C. Alexander, *This New Ocean: A History of Project Mercury* (Washington, DC: National Aeronautics and Space Administration, 1998), 418.

5 John H. Glenn Jr., oral history interview 19, by Jeffrey W. Thomas, May 23, 2008, 4, John Glenn Oral History Project, JGA.

6 "Astronauts' Press Conference," September 16, 1959, National Aeronautics and Space Administration Records, Records of the History Office, Chronological Files, Box 5, September 16–25, 1959 Folder, NARA/FW.

7 James Webb, interview by Martin Weldon, transcript of "Exclusive Space Report," WNEW RADIO, January 21, 1962, 3–4, National Aeronautics and Space Administration Records, Records of the History Office, Chronological Files, Box 20, January–February 1962 Folder, NARA/FW.

8 Kraft, *Flight*, 155.

9 Glenn, "'If You're Shook Up,'" 25.

10 John A. Powers, "National Aeronautics and Space Administration News Briefing," February 13, 1962, 6, National Aeronautics and Space Administration Records, Public Affairs Office, E60, Box 1, Transcript of MA-6 Press Conference 2/20/62 Starlite Palladium Folder, NA/FW.

11 Lyndon B. Johnson, daily diary, January 27, 1962, Lyndon B. Johnson Papers, Vice Presidential, Daily Diary Collection, LBJL, www.lbjlibrary.net/daily-diary.html.

12 Elizabeth Drew, *Campaign Journal: The Political Events of 1983–1984* (New York: Macmillan, 1985), 182.

13 Glenn, oral history interview 19, Thomas, 9.

14 Major Charles Gandy, "National Aeronautics and Space Administration News Briefing," February 13, 1962, 3–4, Box 5, September 16-25, 1959 Folder, NARA/FW.

15 Gandy, "News Briefing," 12.

16 John H. Glenn Jr., "Brief Summary of the MA-6 Orbital Flight," 1, Non-Senate Papers Subgroup, Box 65, Folder 6, JGA.

17 Swenson, Grimwood, and Alexander, *This New Ocean*, 423.

18 John H. Glenn Jr., "MA-6/13 Debriefing, Astronaut Colonel John H. Glenn," February 20 1962, 1–3, National Aeronautics and Space Administration Records, NASA-JSC, Mercury Project Office, E197A, Box 4, Glenn Debrief Folder, NA/FW.

19 Marvin W. Robinson, "Statement for Lieutenant Colonel Glenn in Event of Emergency Landing in Foreign Country," January 12, 1962, 1–2, Non-Senate Papers Subgroup, Box 66, Folder 8, JGA.

20 O. B. Lloyd Jr., "MA-6 Contingencies," January 16, 1962, 2–3, Non-Senate Papers Subgroup, Box 47.1, Folder 23, JGA.

21 Thompson, *Light This Candle*, 319.

22 John H. Glenn Jr., "Pilot's Flight Report," March 1962, 2, National Aeronautics and Space Administration Records, Public Affairs Office, E60, Box 1, Folder MA-6, Pilot Report 2/22/62 Folder, NA/FW.

23 Carpenter later said the words occurred to him as he realized that Glenn would be traveling faster than any human before him. See Carpenter, oral history interview 2, Neal, 52.

24 Wolfe, *Right Stuff*, 251–252.

25 Manned Spacecraft Center and National Aeronautics and Space Administration, *Results of the First United States Manned Orbital Space Flight, February 20, 1962*, 149, https://spaceflight.nasa.gov/outreach/SignificantIncidents/assets/ma-6-results.pdf.

26 Glenn with Taylor, *John Glenn*, 345–346.

27 Glenn, "'If You're Shook Up,'" 28.

28 "NASA Project Mercury Mission MA-6," National Aeronautics and Space Administration, accessed April 15, 2017, http://science.ksc.nasa.gov/history/mercury/ma-6/ma-6.html.

29 Manned Spacecraft Center and National Aeronautics and Space Administration, *Results of the First*, 71.

30 "Testimony of Astronauts and NASA Officials," Senate Committee on Aeronautics and Space Science, February 28, 1962 (Washington, DC: Government Printing Office, 1962), appendix D, "Public Address Announcements by John Powers Beginning at T Minus 22 Minutes Describing MA-6 Launch," 54.

31 Swenson, Grimwood, and Alexander, *This New Ocean*, 426–427.

32 Associated Press, "Around World in 88.29 Minutes—Tremendous," *Pensacola News Journal*, February 21, 1962.

33 Associated Press, "69th Satellite," *San Francisco Examiner*, February 21, 1962.

34 Manned Space Center and National Aeronautics and Space Administration, *Results of the First*, 150.

35 "John Glenn's Earth Orbit Diary from the Friendship 7 in 1962," commemorative webpage, *Newsweek*, February 20, 2017, www.newsweek.com/john-glenn-friendship-7-orbit-earth-diary-558527.

36 Glenn, "Brief Summary of the MA-6," 2, 5.

37 Carpenter with Stoever, *For Spacious Skies*, 279.

38 Dr. Robert B. Voas, "Statements for Foreign Countries," November 6, 1961, National Aeronautics and Space Administration Records, Records of the History Office, Chronological Files, Box 19, November 1961 Folder, NA/FW.

39 Manned Spacecraft Center and National Aeronautics and Space Administration, *Results of the First*, 154.

40 "Glenn Flies Over Kano," *Nigerian Morning Post*, February 21, 1962.

41 Manned Spacecraft Center and National Aeronautics and Space Administration, 1 *Results of the First*, 55.

42 Glenn, oral history interview, Scarborough, 20.

43 Glenn, oral history interview 19, Thomas, 20.

44 "Earth Orbit Diary," *Newsweek.*

45 Swenson, Grimwood, and Alexander, *This New Ocean*, 428.

46 Manned Spacecraft Center and National Aeronautics and Space Administration, *Results of the First*, 156–158.

47 John H. Glenn Jr., "In Orbit," working copy, February 26, 1962, 7–8, Non-Senate Papers Subgroup, Box 66, Folder 9, JGA.

48 Oklahoma Gas and Electric Company, "Thank You! . . . Our Good Friends of Perth," *West Australian*, February 27, 1962.

49 John H. Glenn Jr., "The Mercury-Atlas 6 Space Flight," *Science*, June 1962, 1093.

50 Glenn, "MA-6/13 Debriefing," 55.

51 Glenn, "Mercury-Atlas 6 Space Flight," 1093.

52 Schirra with Billings, *Schirra's Space*, 87.

53 Dr. John O'Keefe, "Report on the Results of the MA-6 Flight in the Field of Space Science," 5, Non-Senate Papers Subgroup, Box 66, Folder 6, JGA.

54 Scott Carpenter, oral history interview, by Michelle Kelly, March 30, 1998, 35, Johnson Space Center Oral History Project.

55 Glenn, "'If You're Shook Up,'" 29.

56 Kluger, "Liftoff," 28.

57 Glenn, "MA-6/13 Debriefing," 8.

58 Glenn, "Brief Summary of the MA-6," 7.

59 Glenn, "Pilot's Flight Report," 8.

60 Glenn, "Pilot's Flight Report," 9–10.

61 Manned Space Center and National Aeronautics and Space Administration, *Results of the First*, 184.

62 Associated Press, "Glenn High above Earth, Kept His Sense of Humor," *Elmira (NY) Advertiser*, February 21, 1962.

63 Kraft, *Flight*, 158.

64 Thompson, *Light This Candle*, 322–323.

65 Thompson, 322.

66 Kraft, *Flight*, 159.

67 Powers, "News Breifing," 4.

68 Kluger, "Liftoff," 30.

69 Manned Space Center and National Aeronautics and Space Administration, *Results of the First*, 188.

70 Glenn, "MA-6/13 Debriefing," 17.

71 Manned Space Center and National Aeronautics and Space Administration, *Results of the First*, 189.

72 Manned Space Center and National Aeronautics and Space Administration, 190.

73 Glenn, "Pilot's Flight Report," 13.

74 Manned Space Center and National Aeronautics and Space Administration, *Results of the First*, 190.

75 Walter Cronkite with John Glenn, "CBS News Special Report: Friendship 7 and John Glenn," Columbia Broadcasting System, April 9, 1962, PCFM.

76 Kluger, "Liftoff," 30.

77 Glenn, "'If You're Shook Up,'" 31.

78 "Forces Deployed around the Globe," *New York Times*, February 21, 1962.

79 Glenn, oral history interview 19, Thomas, 33.

80 Swenson, Grimwood, and Alexander, *This New Ocean*, 433.

81 "Testimony of Astronauts," Senate Committee on Aeronautics and Space Science, 68.

82 Kluger, "Liftoff," 33.

83 Lawrence O'Brien, oral history interview 5, by Michael L. Gillette, December 5, 1985, 1, Lyndon B. Johnson Library Oral History Project, LBJL.

84 Swenson, Grimwood, and Alexander, *This New Ocean*, 433–434.

85 "To Press Pool Randolph from Mittauer," Non-Senate Papers Subgroup, Box 66, Folder 9, JGA.

86 Swenson, Grimwood, and Alexander, *This New Ocean*, 434.

87 United Press International, "Astronaut Was Bit Seasick," *Traverse City (MI) Record-Eagle*, February 21, 1962.

88 Alton Blakeselee, "Astronaut Undergoes Rigid Physical Test," Associated Press, *Greely (CO) Daily Tribune*, February 21, 1962.

89 Glenn, "'If You're Shook Up,'" 31.

90 J. H. Boynton, "Proposed Technical Memorandum: Technical Results of the First Manned Orbital Flight from the United States, Part 1—Mission Results," undated, 4, National Aeronautics and Space Administration Records, Mercury Project Office, MSC Reports, Box 4, Engineering Division Memorandum Re First Manned Orbital Flight Folder, NA/FW.

91 James M. Grimwood, *Project Mercury: A Chronology* (Washington, DC: National Aeronautics and Space Administration, 1963), 158.

92 Carpenter with Stoever, *For Spacious Skies*, 227.

93 Kraft, *Flight*, 160.

94 Because there were so many unknowns and because the US space program had no experience to guide its work, it is surprising that NASA did not lose any lives on an active spacecraft until twenty-four years later when the shuttle *Challenger* accident occurred at liftoff, costing seven lives. In 2003 the *Columbia* burned up in reentry, just as *Friendship 7* might have, and seven more astronauts died. Three Apollo astronauts died during a launchpad test in 1967. In all three accidents, NASA learned valuable lessons that made subsequent flights safer.

95 Edward R. Jones, "Man's Integration into the Mercury Capsule," paper presented at the American Rocket Society's Fourteenth Annual Meeting, November 16–9, 1959, 1, National Aeronautics and Space Administration Records, Records of the History Office, Chronological Files, Box 8, November 6–16, 1959 Folder, NA/FW.

CHAPTER 6: MEANWHILE ON EARTH . . .

1 James Reston, "Halfway to Heaven in Living Color," *New York Times*, February 21, 1962.

2 Cronkite with Glenn, "Friendship 7 and John Glenn."

3 Tom Diemer, "'62 Flight Had Watchers Biting Their Fingernails," *Cleveland Plain Dealer*, October 28, 1998.

4 Gay Talese, "50,000 on Beach Strangely Calm as Rocket Streaks Out of Sight," *New York Times*, February 21, 1962.

5 Associated Press, "People Share Triumph," *Kansas City (MO) Times*, February 21, 1962.

6 Ralph McGill, "All Hearts Beat with Glenn's," *Atlanta Constitution*, February 21, 1962.

7 Saul Pett, Associated Press, "We Pushed with Body and Mind . . . Go, Glenn, Go!" *Lincoln (NE) Star*, February 21, 1962.

8 "Relieved Loop Crowds Extol Glenn as Hero," *Chicago Tribune*, February 21, 1962.

9 Walter Cronkite, "CBS News Extra: The Flight of John Glenn," Columbia Broadcasting System, February 20, 1962, PCFM.

10 "San Bernardinans 'in Orbit' with Astronaut," *San Bernardino County Sun*, February 21, 1962.

11 Arthur Huey, "All Agree: The Flight Was Great!" *Kansas City (MO) Times*, February 21, 1962.

12 Associated Press, "People Share Triumph."

13 "City Rejoices, Prays for Glenn from 'Grandstand' TV Seats," *Philadelphia Inquirer*, February 21, 1962.

14 Associated Press, "Repair Crew Ruins Telecast of Blastoff," *Eau-Claire (WI) Daily Telegram*, February 21, 1962.

15 Charles Moore, "All of Atlanta Watches, Cheers," *Atlanta Constitution*, February 21, 1962.

16 Associated Press, "The John Glenn Haircut Now Offered," *Kansas City (MO) Times*, February 21, 1962.

17 Toni Sherman, "Excitement of Space Triumph Sends Shreveport into Orbit," *Shreveport (LA) Times*, February 21, 1962.

18 Julie Hollabaugh, "There Was One Question, One Prayer Yesterday," *Tennessean* (Nashville), February 21, 1962.

19 Robert F. Hage, "Moral Rights {Responsibilities} of Space Exploration," Women's Guild at Saint Peters Church, February 20, 1962, Non-Senate Papers Subgroup, Box 23, Folder 29, JGA.

20 Robert Wiedrich, "5 Proud Hours Thrill Chicago," *Chicago Tribune*, February 21, 1962.

21 Cronkite, "Flight of John Glenn."

22 "The Day We All Went into Orbit," *New York Mirror*, February 21, 1962.

23 "We All Went into Orbit," *New York Mirror*.

24 Kay Gardella, "Glenn's Orbital Flight A-Okay on Television," *New York Daily News*, February 21, 1962.

25 Cronkite, "Flight of John Glenn."

26 "We All Went Into Orbit," *New York Mirror*.

27 Lawrence O'Kane, "Glenn Fans Upset Decorum at U.N.," *New York Times*, February 21, 1962.

28 Nan Robertson, "New York Pauses to Watch Glenn," *New York Times*, February 21, 1962.

29 "Southlanders Go Wild, Bells Ring Hailing Success of Orbital Flight," *Los Angeles Times*, February 21, 1962.

30 "Flight Echo: TV Watcher Breaks Leg," *Los Angeles Times*, February 21, 1962.

31 "Southlanders Go Wild," *Los Angeles Times*.

32 "Southlanders Go Wild," *Los Angeles Times*.

33 Associated Press, "Cheers Shake Prison Walls at San Quentin," *Long Beach (CA) Independent*, February 21, 1962.

34 Tom Wilson, "Vegas Salutes Glenn, Marine Tribute Here," *Las Vegas Review-Journal*, February 20, 1962.

35 Clayton P. Shepard, "To Colonel John Glenn Jr.," Lyndon B. Johnson Papers, Vice Presidential, 1962 Subject File, Box 182, Science: Space and Aeronautics—Glenn (Orbital Flight) (1 of 2) Folder, LBJL.

36 United Press International, "Glenns Give a Prayer of Thanks," *New York Times*, February 21, 1962.

37 Associated Press, "It's Back to Normal for Glenn Family," *Mount Vernon (IL) Register-News*, February 21, 1962.

38 United Press International, "Never Doubted Success, Say Elder Glenns," *Los Angeles Times*, February 21, 1962.

39 "Thousands in School Follow Orbital Flight," *Los Angeles Times*, February 21, 1962.

40 "Southlanders Go Wild," *Los Angeles Times*.

41 Wilson, "Vegas Salutes Glenn."

42 Lorraine Sakaguchi, letter to John Glenn, undated, Non-Senate Papers Subgroup, Box 32, Folder 46, JGA.

43 "Cincinnati School Pupils 'Take Part' in Glenn Feat," *Cincinnati Enquirer*, February 21, 1962.

44 Wiedrich, "5 Proud Hours."

45 John Glenn, *P.S. I Listened to Your Heart Beat: Letters to John Glenn* (Houston: World Book Encyclopedia Science Service, 1964), 209–210.

46 Mrs. James King, letter to John Glenn, February 23, 1962, Non-Senate Papers Subgroup, Box 23, Folder 39, JGA.

47 Melba Glade, "Children Orbit with Glenn," *Utah Educational Review*, March 1962, 19.

48 "City Rejoices," *Philadelphia Inquirer*.

49 "Man in Space Stamp Issued," *Jefferson City (MO) Daily Capital News*, February 21, 1962.

50 James E. Webb, "Statement on the Issuance of the Mercury Commemorative Stamp," February 20, 1962, Mercury Stamp Folder, Record 00620, HRC.

51 Richard Reeves, *President Kennedy: Profile of Power* (New York: Simon & Schuster, 1993), 285–287.

52 Special to the *New York Times*, "President's Call," *New York Times*, February 21, 1962.

53 Theodore C. Sorensen, *Kennedy* (New York: Harper & Row, 1965), 528–529.

54 Reeves, *President Kennedy*, 287.

55 Associated Press, "Government Stands Still as Glenn Takes Off," *Pensacola Journal*, February 21, 1962.

56 Lyndon B. Johnson, "Vice President's Daily Diary," February 20, 1962, LBJL, www.lbjlibrary.net/collections/daily-diary.html.

57 This is, in many ways, a gesture that runs parallel to LBJ's letters to Caroline and John F. Kennedy Jr. on the night their father was slain. See Lyndon B. Johnson, letter to John David Glenn, February 20, 1962, Lyndon B. Johnson Papers, Vice Presidential, Master File Index, Box 18, Gl-God Folder, LBJL.

58 Russell Porter, "Eyes of Country Turned to Space," *New York Times*, February 21, 1962.

59 Walter A. McDougall, *The Heavens and the Earth: A Political History of the Space Age* (Baltimore: Johns Hopkins University Press, 1997), 347–348.

60 Roger Hilsman, memo to Secretary of State Dean Rusk, March 29, 1962, 1, National Security Files, Box 307, Space Activities General—January 1962–March 1962 Folder, JFKL.

61 United Press International, "Space Committee Member Scores Missile Countdown," *Las Vegas Review-Journal*, February 20, 1962.

62 Cronkite with Glenn, "Friendship 7 and John Glenn."

63 Associated Press, "Aid to Viet-Nam Shows Effect, Says McNamara," *Los Angeles Times*, February 21, 1962.

64 Associated Press, "Kennedy Asks Billions for Federal Pay Raises," *Las Vegas Review-Journal*, February 20, 1962.

65 James D. Cary, "Kennedy Loses Test in Senate," *Abilene (TX) Reporter-News*, February 21, 1962.

66 Associated Press, "For the World to See," *Kansas City (MO) Times*, February 21, 1962.

67 United Press International, "World Cheers Glenn, U.S. Prestige Soars," *San Bernadino-County Sun*, February 21, 1962.

68 Australian Associated Press, "U.S. Astronaut Circles Earth: Sees Lights of Perth on First Orbit," *Sydney Morning Herald*, February 21, 1962.

69 "Launch Heard Here," *Sydney Morning Herald*, February 21, 1962.

70 Peter Retzler, Janos Lazar, Tibor Kozsoker, Kasoly Voros, Miklos Konrad, and Janos Kovacs, letter to John Glenn, undated, Non-Senate Papers Subgroup, Box 28, Folder 64, JGA.

71 Pedro Orozco, letter to John F. Kennedy, February 20, 1962, Non-Senate Papers Subgroup, Box 25, Folder 4, JGA.

72 Truong Thanh Binh, letter to John Glenn, February 22, 1962, Non-Senate Papers Subgroup, Box 28, Folder 63, JGA.

73 "L.A. Officials Praise Frankness of Flight," *Los Angeles Times*, February 21, 1962.

74 "Earth Orbit Diary," *Newsweek*.

75 Editorial, "One of Our Finest Hours," *New York Times*, February 21, 1962.

76 "America Evens the Score and Blazes a Trail," *Battle Creek (MI) Enquirer*, February 21, 1962.

77 McGill, "All Hearts Beat."

78 Cronkite, "Flight of John Glenn."

79 Cynthia Lowry, Associated Press, "3 Major TV Networks Concentrated On Glenn," *Sandusky (OH) Register*, February 21, 1962.

80 Erwin H. Ephron, "40 Million Homes Follow Telecast of First U.S. Orbital Flight," News from Nielsen press release, March 21, 1962, 1–2, Lyndon B. Johnson Papers, Vice Presidential, Box 182, 1962 Subject Files, Science: Space and Aeronautics—Glenn (Orbital Flight) Folder (1 of 2), LBJL.

81 Richard P. Shepard, "$2,000,000 Radio-TV Coverage Carries Story of Flight to Nation," *New York Times*, February 21, 1962.

82 Rick Du Brow, United Press International, "TV Gives Viewers Ringside Seat at Memorable Moment," *Las Vegas Daily Optic* (East Las Vegas, NM), February 21, 1962.

83 Lowry, "3 Major TV Networks."

84 "Identification," *Eugene (OR) Guard*, February 21, 1962.

85 Some analysts have described the four days beginning with JFK's 1963 assassination as the first time TV brought together a national community by providing a way for Americans to figuratively link hands at a time of crisis. However, Glenn's flight truly represents the first time that Americans simultaneously experienced fears and joys minute by minute. Today, the connections forged on that day are easy to forget for two reasons: first, Glenn's flight ended peacefully with no long-lasting negative repercussions; second, many other spaceflights followed it. Kennedy's assassination, on the other hand, has had long-lasting effects and remains a topic for debate within the United States because it has spawned more conspiracy theories than anyone could have imagined in 1963. Glenn's flight was the first hours-long event shared simultaneously by most Americans. It was a true television event: it offered suspense, drama, humor, and a true-blue American hero.

86 To hear Glenn's heartbeat, go to https://youtu.be/lYmc8zZMRnI.

CHAPTER 7: SUPERSTAR

1 Glenn with Taylor, *John Glenn*, 369.

2 Glenn with Taylor, 369.

3 Richard Witkin, "Astronaut and Head of Space Project Get Medals," *New York Times*, February 24, 1962.

4 Gay Talese, "Glenn Family Calm amid Cheers and Confetti," *New York Times*, February 24, 1962.

5 John Glenn, oral history interview, by Walter D. Sohier, June 12, 1964, 4, John F. Kennedy Oral History Collection, JFKL.

6 John Glenn, quoted in John F. Kennedy, "Remarks of the President at the Ceremonies Honoring Astronaut Glenn," February 23, 1962, 2, John F. Kennedy Papers, President's Office Files, Astronauts: John Glenn, JFKPOF-140-020, JFKL.

7 Talese, "Glenn Family Calm," 13.

8 Glenn, oral history interview, Sohier, 3.

9 Carroll Kirkpatrick, "They and Johnson Tour Base Amid Cheering Crowds," *Washington Post*, February 24, 1962.

10 Kennedy, "Honoring Astronaut Glenn," 1.

11 John F. Kennedy, "Text of Citation Awarded Lieutenant Colonel John H. Glenn Jr., United States Marine Corps," February 23, 1962, 1, John F. Kennedy Papers, Presidential Office Files, Astronauts: John Glenn, JFKPOF-095-023, JFKL.

12 The irony of this comment is that the 1960s date that would become most indelibly imprinted on the American psyche was not the date of a spaceflight, but November 22, 1963—the day Kennedy himself was assassinated in Dallas. Furthermore, Kennedy was clearly being diplomatic because the dates of Shepard's and Grissom's flights never became cultural touchstones. See Kennedy, "Honoring Astronaut Glenn."

13 John Glenn, "Press Conference, Flight of MA-6," February 23, 1962, 2, National Aeronautics and Space Administration Records, E60, Public Affairs Office, Box 1, Press Conference Flight of MA-6 Folder, NA/FW.

14 John G. Norris, "Astronaut Cites Fears over Possible Loss of Heat-Shield," *Washington Post*, February 24, 1962.

15 Rev. Lloyd George Schell, "The Faith of an Astronaut: Lt. Col. John H. Glenn Jr., USMC," February 25, 1962, 1–3, Non-Senate Papers Subgroup, Box 3, Folder 32, JGA.

16 "Glenn Takes Nation by Storm—Including Congress," *Space News Roundup!*, March 7, 1962, 1.

17 "Astronaut John Glenn's Washington, D.C., Parade," United States National Aeronautics and Space Administration Films, February 26, 1962, United States Government Agencies Collection, Reel 5, JFKL.

18 It is only a joint *session* of Congress if the two legislative bodies are called together to conduct official business, such as hearing a president's State of the Union address. Heads of state and people like Douglas MacArthur and John Glenn visit a meeting.

19 Dora Jane Hamblin, "Applause, Tears and Laughter and the Emotions of a Long-Ago Fourth of July," *Life*, March 9, 1962, 34–35.

20 Excerpt from *Congressional Record*, February 26, 1962, Non-Senate Papers Subgroup, Box 48.1, Folder 2, JGA.

21 Wilfrid C. Rodgers, "'My Happiest Day,'" *Boston Globe*, February 27, 1962.

22 Shannon Stirone, "The Real Cost of NASA Missions," *Popular Science*, November 4, 2015, www.popsci.com/real-cost-nasa-missions#page-2.

23 *Orbital Flight of John H. Glenn Jr.: Hearing Before the Senate Committee on Aeronautical and Space Sciences, February 28, 1962* (Washington, DC: Government Printing Office, 1962), 13.

24 McCandlish Phillips, "Vast Turnout for Glenn; Parade in Paper Blizzard 'Overwhelms' Astronaut," *New York Times*, March 2, 1962.

25 "Heroes: 'I Touched Him!'" *Newsweek*, March 12, 1962, 25.

26 "'I Touched Him!'" *Newsweek*, 25.

27 Harry Reasoner, "CBS News Extra: John Glenn in New York," Columbia Broadcasting Company, March 1, 1962, PCFM.

28 Koppel, *Astronaut Wives Club*, 79.

29 "Youths Boom 'Glenn 4 President,'" *New York Times*, March 2, 1962.

30 Nan Robertson, "Screaming Teen-Agers Set Tone for Welcome to Astronauts," *New York Times*, March 2, 1962.

31 "'I Touched Him!'" *Newsweek*, 25.

32 Reasoner, "John Glenn in New York."

33 Reasoner, "John Glenn in New York."

34 Nikita Khrushchev, telegram to John F. Kennedy, February 21, 1962, 1, John F. Kennedy Papers, Presidential Office Files, Astronauts: John Glenn, JFKPOF-095-024, JFKL.

35 Sam Pope Brewer, "An Informal U.N. Acclaims Glenn," *New York Times*, March 3, 1962.

36 "That Persistent Rider on Glenn's Shirttails," *New York Herald Tribune*, March 2, 1962, Lyndon B. Johnson Papers, Vice Presidential, 1962 Subject File, Box 182, Science: Space and Aeronautics—Glenn (Orbital Flight) Folder (2 of 2), LBJL.

37 See multiple letters and telegrams, Lyndon B. Johnson Papers, Vice Presidential, 1962 Subject File, Box 182, Science: Space and Aeronautics—Glenn (Orbital Flight) Folder (1 of 2), LBJL.

38 Glenn, oral history interview, Sohier, 18.

39 O'Brien, oral history interview 5, Gillette, 2.

40 Peter Kihss, "All 95 on Jetliner Killed in Crash into Bay on Take-Off at Idlewild; President Spurs Federal Inquiry," *New York Times*, March 2, 1962.

41 The highest US death toll in a plane crash to date—134—occurred when two airliners collided over Staten Island in 1960. Another two-plane disaster occurred over the Grand Canyon in 1956, taking 128 lives. See "Crash Was 5th Worst in History and Worst for One U.S. Airliner," *New York Times*, March 2, 1962.

42 Alan Gonder, "Glenn Greeted By 75,000 on Visit to New Concord," *Zanesville (OH) Sunday Times-Recorder*, March 4, 1962.

43 Mrs. John H. Glenn Sr., "The World My Son Took with Him," *Guideposts*, January 1963, 3.

44 "And Then the Band Played 'When Johnny Comes Marching Home...'" *Life*, March 9, 1962, 36–37.

45 "Colonel Wonderful," *Time*, March 9, 1962, 22.

46 "Meet Orbit Hill," *Life*, March 9, 1962, 2–3.

47 Glenn, *I Listened to Your Heart Beat*, 42.

48 Wolfe, *Right Stuff*, 318.

49 "A Snow Replica of a Famous Capsule," Non-Senate Papers Subgroup, Box 3, Folder 32, JGA.

50 Lewis Cutrer, telegram to Lyndon B. Johnson, February 22, 1962, Lyndon B. Johnson Papers, Vice Presidential, 1962 Subject File, Box 182, Science: Space and Aeronautics—Glenn (Orbital Flight) Folder (1 of 2), LBJL.

51 "Capsule, Mercury, MA-6," Smithsonian Learning Lab, https://learninglab.si.edu/resources/view/23194#more-info.

52 Glenn, *I Listened to Your Heart Beat*, ix–x.

53 Glenn with Taylor, *John Glenn*, 405.

54 Glenn, *I Listened to Your Heart Beat*, 69.

55 National Aeronautics and Space Administration, "Summary of Mail Processed for Lt. Col. John H. Glenn Jr., February 20 thru August 31, 1962," 1, Non-Senate Papers Subgroup, Box 23, Folder 1, JGA.

56 National Aeronautics and Space Administration, "Summary of Mail," 2.

57 Foreign Broadcast Information Service, *Daily Report Supplement, World Reaction Series, No. 3—1962*, February 23, 1962, 11, Non-Senate Papers Subgroup, Box 65, Folder 23, JGA.

58 Glenn, *I Listened to Your Heart Beat*, xi.

59 *Heart Beat*, 7.

60 *Heart Beat*, 25.

61 *Heart Beat*, 84.

62 *Heart Beat*, 248.

63 Chet Hagan, letter to Annie Glenn, March 7, 1962, Non-Senate Papers Subgroup, Box 23, Folder 36, JGA.

64 "Request for Appearances: John H. Glenn Jr.," Non-Senate Papers Subgroup, Box 19, Folder 18, JGA.

65 *The Nation's Future*, transcript, moderated by Edwin Newman, National Broadcasting Company, May 6, 1962, 20, Non-Senate Papers Subgroup, Box 47.1, Folder 50, JGA.

66 Glenn with Taylor, *John Glenn*, 389.

67 "Astronaut Glenn Says Nation Needs Scout-Trained Leaders," *"Go" Roundup News*, Boy Scouts of America, Fall 1962, New Brunswick, New Jersey, 1, Non-Senate Papers Subgroup, Box 47, Folder 18, JGA.

68 Vincent Butler, "4 Million Go Wild in N.Y., for Col. Glenn," *Chicago Tribune*, March 2, 1962.

69 J. A. Brunton Jr., letter to John H. Glenn Jr., April 12, 1962, Non-Senate Papers Subgroup, Box 98.1, Folder 3, JGA.

70 "Astronaut Glenn Says," *"Go" Roundup News*, 1.

71 Adlai Stevenson, "Address to the 12th Annual Conference of National Organizations, American Association for the Support of the United Nations," March 13, 1962, Non-Senate Papers Subgroup, Box 19, Folder 10, JGA.

72 Cronkite with Glenn, "Friendship 7 and John Glenn."

73 See letters in Lyndon B. Johnson Papers, Vice Presidential, 1962 Subject File, Box 182, Science: Space and Aeronautics—Glenn (Orbital Flight) Folder (1 of 2), and Science: Space and Aeronautics Folder, LBJL.

74 Even at the time of his death, a *Forbes* online article condemned his refusal to encourage NASA to find a way to admit women to the astronaut corps sooner. See Meta S. Brown, "How John Glenn Thwarted Female Astronauts and Why It Still Matters for Minorities and Women in Tech," *Forbes*, December 19, 2016, www.forbes.com/sites/metabrown/2016/12/19/how-john-glenn-thwarted-female-astronauts-and-why-it-still-it-for-matters-minorities-and-women-in- tech/#1231387e2be6.

75 John Glenn, Testimony before the Special Subcommittee on the Selection of Astronauts of the House of Representatives Committee on Science and Astronautics, July 18, 1962 (Washington, DC: Government Printing Office, 1962), 64.

76 Glenn, Testimony, 69.

77 Official NASA Memo about Glenn Flown Currency," Jefferson Space Museum Blog, January 26, 2008, http://jeffersonspacemuseum.com/jefferson-in-space/2009/10/official-nasa-memo-about-glenn-flown.html.

78 "Item 6087—Guenter Wendt's Mercury-Atlas 6 Friendship 7 Flown Dollar Bill," RR Auction Past Auction Item, April 2016, www.rrauction.com/PastAuctionItem/3354700.

79 McDougall, *Heavens and the Earth*, 347.

CHAPTER 8: WHAT'S NEXT?

1 Although this secret document was dated January 1962, it did not reach Kennedy's national security adviser, McGeorge Bundy, until May. See National Aeronautics and Space Administration, "The Long Range Plan," January 1962, 1–2, Space Activities: Long Range Plans of NASA, Vol. 1–4 Folder, JFKL, www.jfklibrary.org/asset-viewer/archives/JFKNSF/307/JFKNSF-307-001.

2 Edward C. Welsh, "United States Policy on Outer Space," undated, 2, Space Activities: General, Vol. 1–3, 1962: June–July Folder, JFKL, www.jfklibrary.org/asset-viewer/archives/JFKNSF/307/JFKNSF-307-008.

3 Welsh, "Policy on Outer Space," 3.

4 John F. Kennedy, "National Security Action Memorandum 156," May 26, 1962, National Security Files, Box 339 NSAM 156 Folder, JFKL.

5 Glenn, oral history interview, Sohier, 18–19.

6 *The Nation's Future*, transcript, National Broadcasting Company, May 4, 1962, 8, Non-Senate Papers Subgroup, Box 47.1, Folder 50, JGA.

7 *Nation's Future*, transcript, 6–7.

8 *Nation's Future*, 13. Interestingly, experiments during Glenn's 1998 flight on the space shuttle *Discovery* focused on the issue of the effects of weightlessness.

9 *Nation's Future*, 24–25.

10 Titov's reply may have been entirely accurate, but because Americans viewed the Soviet government as being incredibly secretive, many suspected that space failures could be hidden in these early years before American satellites were in place to monitor such happenings. When Vladimir Komarov died in a 1967 space mission, the Soviet Union acknowledged what had happened, as it did when three more cosmonauts died unexpectedly at the end of a mission in 1971.

11 *Nation's Future*, 35–37.

12 *Nation's Future*, 30.

13 *Nation's Future*, 38.

14 Slava Gerovitch, *Soviet Space Mythologies* (Pittsburgh: University of Pittsburgh Press, 2015), 147.

15 John A. Powers and C. Oleo DeOrsey, news conference, April 3, 1962, 2, Record 207, HRC.

16 Wolfe, *Right Stuff*, 285.

17 Koppel, *Astronaut Wives Club*, 99.

18 Glenn, oral history interview, Sohier, 7.

19 United Press International, "Glenn, Space Bosses Deny Schirra Charges," *Boston Record American*, September 15, 1962, Glenn, John H. Jr., NASA Consultant, Former Astronaut Folder, HRC.

20 Wolfe, *Right Stuff*, 313–314.

21 Kraft, *Flight*, 164–170.

22 "NASA-Telstar Contract," news release, National Aeronautics and Space Administration, July 8, 1962, National Security Files, Box 282, NASA 1962 Folder, JFKL.

23 John F. Kennedy, taped meeting conversation on space budget, "Meetings: Tape 63. Space Program Budget, Plans and Priorities," November 21, 1962, JFKPOF-MTG-062, JFKL.

24 Kraft, *Flight*, 181–183.

25 Shepard and Slayton with Barbree, *Moon Shot*, 149.

26 John H. Glenn Jr., oral history interview 20, by Jeffrey W. Thomas, October 20, 2008, 3–4, John Glenn Oral History Project, JGA.

27 Shepard and Slayton with Barbree, *Moon Shot*, 149.

28 United Press International, "Glenn Picked to Pioneer Moonshot; Duties Assigned All 16 Astronauts," January 27, 1963, *Washington Post*, reprinted in *NASA Current News*, Glenn, John H. Jr. (1962–63) Folder, Record 7710. HRC.

29 Glenn, oral history interview 20, Thomas, 2.

30 Wolfe, *Right Stuff*, 308–309.

31 National Aeronautics and Space Administration, "Astronaut Personal Appearances," Records of NASA, SP27.1, Box 42, Personal Appearances Jan.–Aug. 1963 Folder, NA/FW.

32 Mrs. John H. Glenn Sr., "World My Son Took," 1.

33 John H. Glenn Jr., "Glenn Looks at Life in U.S.," *Atlanta Journal and Constitution*, April 28, 1963, Glenn, John H. Jr., (1962–63) Folder, Record 7710, HRC.

34 Lt. Col. J. H. Glenn Jr., "Book Larnin' Ain't Everything," *Utah Educational Review*, May 1963, 37.

35 Ken Hechler, "John Glenn and the American Dream," *Congressional Record—Appendix*, September 16, 1963, A5798.

36 Steven R. Derounian, "New Frontier Astronaut?" *Congressional Record—Appendix*, October 23, 1963, A6617.

37 "Astronauts' Million-Dollar Deal: Pro and Con of Argument over Spacemen's Stories," *U.S. News & World Report*, September 30, 1963, 8–9.

38 Lt. Col. John Glenn, "Why Astronauts Sold Their Personal Stories: Interview with Lt. Col. John Glenn," *U.S. News & World Report*, September 30, 1963, 9–10.

39 John F. Kennedy and James Webb, taped conversation, "Meetings: Tape 111," Lunar Program (James Webb), September 18, 1963, JFKL.

40 Doug Freelander, "16 Astronauts Leave for Panama Jungle Training," Non-Senate Papers Subgroup, Box 69, Folder 40, JGA.

41 Glenn, oral history interview 18, Thomas, 46–47.

42 Glenn with Taylor, *John Glenn*, 391–392.

43 Lt. Col. John H. Glenn Jr., proposal to McGeorge Bundy concerning spaceflight information negotiations with the Russians, November 4, 1963, Non-Senate Papers Subgroup, Box 64, Folder 43, JGA.

44 Lt. Col. John H. Glenn Jr., proposal to McGeorge Bundy concerning spaceflight international scientific briefings, November 4, 1963, Non-Senate Papers Subgroup, Box 64, Folder 43, JGA.

45 James E. Webb, memorandum to Dr. Hugh L. Dryden and Dr. Robert C. Seamans, November 8, 1963, James Webb Papers, Box 4, November 1963 Folder, JFKL.

46 Glenn, oral history interview, Scarborough, 21–22.

47 The issue of the long-range effects of weightlessness helped to prompt Glenn's second flight in 1998. See Glenn, oral history interview, Sohier, 16.

48 John Glenn, interview by Robert Fabry, October 5, 2009, 3, Non-Senate Papers Subgroup, Box 110, Folder 98, JGA.

49 Koppel, *Astronaut Wives Club*, 120–121.

50 John Glenn, oral history interview 1, by Roberta Greene, June 26, 1969, 2, Robert F. Kennedy Oral History Project, JFKL.

51 Prohibitions against political campaigning by members of the military were mandated by Department of Defense Directive 1344.10, titled "Political Activities by Members of the Armed Forces." This directive is similar to the Hatch Act, approved by Congress in 1939, which prohibits political activities by federal government employees.

52 Lyndon Johnson and Wayne Hays, telephone conversation, January 15, 1964, Discover LBJ, www.discoverlbj.org/item/tel-01378.

53 Glenn, oral history interview 20, Thomas, 20.

54 Glenn, oral history interview 20, Thomas, 22.
55 Wallace M. Greene Jr., letter to James Webb, January 22, 1964, NASA Records, Microfilm Reel 2, JFKL.
56 Glenn, oral history interview 20, Thomas, 26.
57 Glenn with Taylor, *John Glenn*, 403–404.
58 James Reston and John Glenn, interview transcription, November 10, 1964, 100, Non-Senate Papers Subgroup, Box 98.1, Folder 22, JGA.
59 Thompson, *Light This Candle*, 355.
60 Cunningham, *All-American Boys*, 61.
61 John H. Glenn Jr., oral history interview 21, by Jeffrey W. Thomas, 3, January 13, 2009, 3, John Glenn Oral History Project, JGA.
62 W. H. Glenn and John Glenn, signed Royal Crown Cola contract, August 6, 1964, Non-Senate Papers Subgroup, Box 97.1, Folder 33, JGA.
63 United Press International, untitled, October 24, 1966, Non-Senate Papers Subgroup, Box 97.1, Folder 25, JGA.
64 Royal Crown Cola, "'Booming World Market' Target for 'New' Royal Crown Cola International Ltd.," press release, October 24, 1966, Non-Senate Papers Subgroup, Box 97.1, Folder 25, JGA.
65 Glenn, oral history interview 21, Thomas, 4.
66 Dale Butland, "John Glenn: The Last American Hero?" *New York Times*, December 10, 2016.
67 R. W. Hale, memorandum to Edward Welsh, June 8, 1966, Lyndon B. Johnson Papers, Outer Space, OS2, Box 75, OS4-1, Astronauts Folder, LBJL.
68 Ridgway B. Knight, note to State Department, May 23, 1966, Lyndon B. Johnson Papers, Confidential File: Outer Space, Box OS4, OS 4-1 Astronauts 9/16/65–6/8/66 Folder, LBJL.
69 Associated Press, "French Give Spaceman Glenn a Wild Greeting," *Miami Herald*, May 28, 1966, Glenn, John H. Jr. (1966–70) Folder, Record 802, HRC.
70 New Nine astronauts Elliot See and Charles Bassett were killed in a NASA trainer jet accident in February 1966 on the way to practice space rendezvous exercises in preparation for their scheduled Gemini 9 flight. Pilot error was blamed for their deaths.
71 Lyndon Johnson and James Webb, telephone conversation #11632, March 10, 1967, Discover LBJ, www.discoverlbj.org/item/tel-11632.
72 Don James, "John Glenn on Foot in Africa," *Steelways*, 2–4, Non-Senate Papers Subgroup, Box 98, Folder 8, JGA.
73 "Trail of Stanley and Livingstone," revised script, November 30, 1967, Non-Senate Papers Subgroup, Box 98, Folder 25, JGA.
74 Glenn, oral history interview 21, Thomas, 13–16.
75 Dave Glenn, journal, June 1967, Non-Senate Papers Subgroup, Box 98, Folder 23, JGA.
76 Glenn, oral history interview 21, Thomas, 16.

CHAPTER 9: GLENN'S FRIEND, ROBERT KENNEDY

1 Glenn with Taylor, *John Glenn*, 425–426.
2 Glenn, oral history interview, Sohier, 1–3.
3 John Glenn, untitled statement, 7, Senate Papers Subgroup, Box 53, Folder 33, JGA.
4 Glenn with Taylor, *John Glenn*, 385.
5 Glenn, oral history interview 1, Greene, 6.
6 "Around the World," *Dover (OH) Daily Reporter*, February 26, 1968.
7 Glenn, untitled statement, 7.
8 His last daughter, Rory, was born in December 1968, six months after her father's death.
9 John Glenn, filed version of statement for Memorial Book, 6, Senate Papers Subgroup, Box 53, Folder 33, JGA.
10 James W. Hilty, *Robert Kennedy: Brother Protector* (Philadelphia: Temple University Press, 1997), 20.

11 Glenn, untitled statement, 9.

12 Glenn, untitled statement, 8.

13 Glenn, untitled statement, 1.

14 See John H. Glenn Jr., "The Age of Imagination and Inquiry," undated, Non-Senate Papers Subgroup, Box 47.1, Folder 15, JGA.

15 Glenn, statement for Memorial Book, 7.

16 Glenn, untitled statement, 3.

17 Glenn, statement for Memorial Book, 2.

18 Glenn, statement for Memorial Book, 6.

19 Glenn, oral history interview 1, Greene, 2.

20 Glenn, untitled statement, 10–11.

21 Glenn, oral history interview 1, Greene, 6.

22 Robert Kennedy had supported President Kennedy's expansion of the number of US military "advisor" troops in South Vietnam to 16,300 by 1963. However, he strongly opposed Johnson's policy that sent more than half a million American men into the warzone.

23 Glenn, oral history interview 1, Greene, 4.

24 John Glenn, oral history interview 2, by Roberta Greene, June 30, 1969, 22, Robert F. Kennedy Oral History Project, JFKL.

25 Glenn, oral history interview 1, Greene, 4.

26 Evan Thomas, *Robert Kennedy: His Life* (New York: Simon & Schuster, 2000), 62.

27 James R. Jones, note, March 21, 1968, Lyndon B. Johnson Papers, White House Central Files, Name File, Box G144, Glenn, John Folder, LBJL.

28 Al Fox, "Kennedy Praises UA, Dr. Rose; Col. Glenn a 'Bonus' for Students," *Birmingham News*, March 22, 1968.

29 Glenn, oral history interview 1, Greene, 9.

30 Glenn with Taylor, *John Glenn*, 423–424.

31 Arthur M. Schlesinger Jr., *Robert Kennedy and His Times* (Boston: Houghton Mifflin, 1978), 874.

32 Glenn, oral history interview 1, Greene, 13.

33 Alice George, "When Robert Kennedy Delivered the News of Martin Luther King's Assassination," *Smithsonian Magazine*, April 2, 2018, www.smithsonianmag.com /smithsonian-institution/emotionally-wounded-robert-kennedy-delivers-news-kings -assassination-180968625.

34 "Robert Kennedy: Delivering News of King's Death," NPR, April 4, 2008, www.npr .org/2008/04/04/89365887/robert-kennedy-delivering-news-of-kings-death.

35 C. William Ashley, "Glenn, Hilsman Recruit RFK Supporters," *Ohio State Lantern*, April 25, 1968.

36 Nebraskans for Kennedy, "John Glenn's Nebraska Schedule—Saturday, May 11, 1968," Senate Papers Subgroup, Box 54, Folder 4, JGA.

37 "Colonel John Glenn's Schedule," June 3, 1968, Senate Papers Subgroup, Box 53, Folder 30, JGA.

38 Glenn, untitled statement, 6.

39 Douglas Perry, "Robert Kennedy, Eugene McCarthy Thrilled Young Battles During Epic Battle for 1968 Oregon Primary," *Oregonian* (Portland), May 16, 2016, www.oregonlive .com/history/2016/05/robert_kennedy_eugene_mccarthy.html.

40 Glenn, statement for Memorial Book, 2.

41 Schlesinger, *Robert Kennedy*, 908.

42 Glenn, oral history interview 2, Greene, 17.

43 Glenn, statement for Memorial Book, 4.

44 Glenn, oral history interview 2, Greene, 22–23.

45 Glenn, oral history interview 2, Greene, 18.

46 Glenn, oral history interview 1, Greene, 10.

47 Glenn, oral history interview 1, Greene, 14.

48 Glenn, oral history interview 1, Greene, 11.

49 Hodding Carter Jr., oral history interview, by T. H. Baker, November 8, 1968, 25–26, Lyndon B. Johnson Oral History Collection, LBJL.

50 Robert J. Herguth, "Gun Control Gets Boost as Store Bans Sale of Toy Firearms," *Chicago Daily News*, June 7, 1968.

51 Phyllis Battelle, "Assignment: America, Famed Families Share Faith," *Piqua (OH) Daily Call*, June 24, 1968.

52 Glenn, oral history interview 2, Greene, 24–25.

53 "Notes Re RFK Funeral Arrangements," undated, Senate Papers Subgroup, Box 53, Folder 30, JGA.

54 "Memorandum for: Pallbearers," Senate Papers Subgroup, Box 53, Folder 30, JGA.

55 Glenn, unititled statement, 4.

56 Russell Baker, "Observer: On the Southbound Train," *New York Times*, June 11, 1968.

57 George, "Kennedy Delivered the News."

58 United Press International, "John Glenn Vocal Supporter for Strong Gun Control Legislation," *Lancaster (OH) Eagle-Gazette*, June 27, 1968.

59 "Across the Nation," *Akron Beacon Journal*, August 4, 1968.

60 Ken Gookins, "In the Great Outdoors," *Zanesville (OH) Times Recorder*, August 18, 1968.

61 James R. Jones, telegram to John Glenn et al., October 21, 1968, Lyndon B. Johnson Papers,

62 Butland, interview by author.

CHAPTER 10: INTRODUCTION TO POLITICAL LIFE

1 John H. Glenn Jr., "Pre-Apollo 10," 6–9, Non-Senate Papers Subgroup, Box 47.1, Folder 20, JGA.

2 John Glenn, untitled article, 1–3, Non-Senate Papers Subgroup, Box 47, 1, Folder 18, JGA.

3 John H. Glenn Jr., "Where Do We Go from Here in Space?" World Book Science Service, 1970, 2–5, Non-Senate Papers Subgroup, Box 471, Folder 21, JGA.

4 "U.S. Senator Stephen M. Young's Television and Press Conference," February 24, 1970, 1, Senate Papers Subgroup, Box 136, Folder 34, JGA.

5 Donald Janson, "Glenn Is Opposed in Senate Contest," *New York Times*, December 18, 1969.

6 Glenn with Taylor, *John Glenn*, 428.

7 Ohio Press Service, untitled, March 13, 1970, Senate Papers Subgroup, Box 136, Folder 14.

8 See Justin Moldow, "John Glenn Is No American Hero. He Voted for Gun Control. Twice," Liberty Hangout, December 8, 2016, http://libertyhangout.org/2016/12/john-glenn-is-no-american-hero-he-voted-for-gun-control-twice.

9 O'Brien, oral history interview 5, Gillette, 1–2.

10 "Second Cadell Poll," Senate Papers Subgroup, Box 136, Folder 14, JGA.

11 R. W. Apple Jr., "Glenn Depressed by Student Riots," *New York Times*, May 5, 1970.

12 Daniel Strohl, "Infamous 1971 Dodge Challenger to Take Part in Largest-Ever Display of Indy 500 Pace Cars," *Hemmings Daily*, October 20, 2014, www.hemmings.com/blog/2014/10/20/infamous-1971-dodge-challenger-to-take-part-in-largest-ever-display-of-indy-500-pace-cars.

13 Travis M. Andrews, "Annie Glenn: 'When I Called John, He Cried. People Just Couldn't Believe That I Could Really Talk,'" *Washington Post*, December 9, 2016, www.washingtonpost.com/news/morning-mix/wp/2016/12/09/to-john-glenn-the-real-hero-was-his-wife-annie-conqueror-of-disability/?utm_term=.dc9cdfe5af1c.

14 John Glenn, "Annie Glenn," My Hero Stories, August 11, 2014, https://myhero.com/Annie_Glenn_bk06. In fact, Lyn once stepped on a nail, and Annie was unable to call 911, so she had to dash to find a neighbor to make the call for help.

15 "What Annie Glenn Means to ASHA and Our Members," YouTube video, posted by "American Speech-Language- Hearing Association," August 6, 2015, https://youtu.be/vusErfa-Sns.

16 Glenn with Taylor, *John Glenn*, 432–433.

17 Glenn with Taylor, 434.

18 Joseph D. Rice, "Metzenbaum Spirits Lifted by $100 Diners," *Cleveland Plain Dealer*, April 19, 1974.

19 Robert G. McGruder, "500 Join Glenn for 99-Cent Meal at Flat Iron Café," *Cleveland Plain Dealer*, April 19, 1974.

20 "News from Citizens for John Glenn," February 18, 1974, 1, Senate Papers Subgroup, Box 141, Folder 12, JGA.

21 "News from Citizens for John Glenn," 1.

22 Christopher Lydon, "Rematch of Glenn and Metzenbaum in Ohio Primary: Controversy on Tax Returns Reflects Watergate Impact," *New York Times*, April 21, 1974.

23 Arlen J. Large, "For Sen. Metzenbaum, Being an Incumbent Has Its Negative Side," *Wall Street Journal*, April 18, 1974.

24 Brian T. Usher, "Glenn Claims Rival Bared Tax Returns Out of Desperation," *Cleveland Plain Dealer*, April 25, 1974.

25 "John Glenn: Gold Star Mothers Speech 1974," YouTube video, posted by "jwat072," October 26, 2017, https://youtu.be/0tjmZPRN-i8.

26 "John Glenn: Gold Star Mothers Speech," YouTube video, posted by "WOSU Public Media," December 9, 2016, https://youtu.be/a3gThGnavc0.

27 Myra MacPherson, "The Hero-as-Politician: John Glenn's Hard Road," *Washington Post*, January 12, 1975, Potomac section—1, Glenn, John H. Jr., NASA consultant, Former Astronaut Folder, Record 13310, HRC.

28 "Rookies of the Year," *Newsweek*, November 18, 1974, 28, Glenn, John H. Jr. Folder, Record 7712, HRC.

29 David Hafemeister, interview by author, January 4, 2018.

30 Connolly, interview by author.

31 Zeke J. Miller, "A Life of Public Service," in Gibbs, "John Glenn," 49.

32 "John Herschel Glenn—Junior Senator from Ohio," undated, 3–4, Carter/Mondale Campaign Committee, Vice Presidential Candidates, Box 96, JCL.

33 Miller, "Life of Public Service," 50.

34 Spencer Rich, "Gas Rationing Power Voted," *Washington Post*, April 11, 1975.

35 Philip Shabecoff, "Ford Signs Bill on Energy that Ends Policy Impasse and Cuts Crude Oil Prices," *New York Times*, December 23, 1975.

36 Marjorie Hunter, "An Astronaut Quietly Makes Mark in Senate," *New York Times*, February 4, 1976.

37 "John Herschel Glenn—Junior Senator from Ohio," 3.

38 "Vice Presidential Choices and Acceptability by American Jewish Community," June 1976, Carter/Mondale Campaign Committee, Vice Presidential Candidates, Box 96, JCL.

39 Stu Eizenstat, memorandum to Elliott Weiss, July 1, 1976, Carter/Mondale Campaign Committee, Vice Presidential Candidates, Box 96, JCL.

40 Steve Travis, memorandum to Stuart Eizenstat, July 6, 1976, Carter/Mondale Campaign Committee, Vice Presidential Candidates, Box 96, JCL.

41 Robert E. Kintner, memorandum to Hamilton Jordan, July 9, 1976, 6, Records of the Office of the Staff Secretary, 1976–1981, 1976 Campaign Transition File, 1977–1977, Campaign- Miscellaneous, 6/76–10/76, National Archives Identifier: 148843, JCL.

42 Joseph D. Rice, "Carter's Advisers Split on VP Choice," *Cleveland Plain Dealer*, July 12, 1976.

43 Walter Cronkite, *Conversations with Cronkite* (Austin: Dolph Briscoe Center for American History, 2010), 318.

44 R. W. Apple Jr., "Upbeat Convention Is Ready to Nominate Carter Tomorrow," *New York Times*, July 13, 1976.

45 John Glenn, "Keynote Speech of Senator John Glenn, Democratic National Convention," July 12, 1976, 5, Senate Papers Subgroup, Box 50, Folder 24, JGA.

46 Patrick Buchanan, "Mondale the Spender a Loser," *Chicago Tribune*, July 20, 1976.

47 John H. Glenn Jr., oral history interview 23, by Jeffrey W. Thomas, June 19, 2009, 2–3, John Oral History Project, JGA.

48 Previously House minority leader, Ford was chosen by Nixon and approved by Congress to fill the vacancy created by Spiro Agnew's resignation as vice president. As part of a plea deal, Agnew pleaded nolo contendere to a single tax evasion charge. He had faced a slew of extortion, bribery, and tax evasion allegations.

49 "Bills Sponsored by John Glenn 95th Congress," undated, 1, Senate Papers Subgroup, Box 435, Folder 1, JGA.

50 Christine McCreary, oral history interview, by Donald A. Richie, May 19, 1998, 30–33, United States History Office—Oral History Project.

51 Connolly, interview by author.

52 Butland, interview by author.

53 John H. Glenn Jr., oral history interview 22, by Jeffrey W. Thomas, February 20, 2009, 13–14, John Glenn Oral History Project, JGA.

54 "The Nuclear Non-Proliferation Act of 1978 Should Be Selectively Modified," summary of report, US General Accounting Office, May 21, 1981, www.gao.gov/products/115322.

55 Glenn, oral history interview 22, Thomas, 14–15.

56 John Glenn, remarks before the American Defense Preparedness Association's annual luncheon, May 17, 1979, 1–2, Senate Papers Subgroup, Box 60, Folder 57, JGA.

57 Glenn with Taylor, *John Glenn*, 451.

58 "Milestones: 1969–1976; Strategic Arms Limitations Talks/Treaty (SALT) I and II," US Department of State Office of the Historian, accessed June 26, 2019, https://history.state.gov/milestones/1969-1976/salt.

59 Robert G. Kaiser, "Sen. Glenn Won't Vote for SALT in Committee," *Washington Post*, November 8, 1979.

60 Dan Balz, "Methodical Hero with a Zest for Duty," *Washington Post*, January 15, 1984.

61 Charles Mohr, "Senate Panel Issues Arms Pact Report, Urging Its Approval," *New York Times*, November 20, 1979.

62 Richard Burt, "Senate Panel Votes Antitreaty Report," *New York Times*, December 21, 1979.

63 "Strategic Arms Limitation Treaty II (1979)," atomicarchive.com, accessed January 8, 2019, www.atomicarchive.com/Treaties/Treaty13.shtml.

64 Zbigniev Brzezinski, memorandum to President Jimmy Carter, undated, 1, White House Central Files, National Security Defense: ND-7 (Confidential), JCL.

65 The number of US troops in South Korea declined by only 333 from the 1976 level that existed before Carter took office to the level during 1980, his last full year in office. See "Number of U.S. troops in South Korea," NZ-DPRKSociety, accessed January 1, 2019, https://sites.google.com/site/nzdprksociety/number-of-us-troops-in-rok-by-year.

66 "B-1B Lancer," Boeing, accessed January 1, 2019, www.boeing.com/defense/b-1b-bomber.

67 Glenn, oral history interview 22, Thomas, 26–35.

68 "Verbatim Transcript of Meeting with People's Republic of China Vice Premier Deng Xiaoping, Great Hall, Peking, China," January 9, 1979, 2, Senate Papers Subgroup, Box 57, Folder 2, JGA.

69 Hedrick Smith, "President Concedes," *New York Times*, November 5, 1980.

70 In 1992, President George H. W. Bush joined the ranks of presidents who had been elected to a first term and lost their reelection bids. During the twentieth century, Taft, Hoover, Carter, and Bush were the only presidents to encounter that fate. Ford was an incumbent who lost reelection, but he had been appointed vice president and ascended to the presidency without ever being elected.

71 "Statewide Summary Sheet—November 4, 1980 General Election," Senate Papers Subgroup, Box 149, Folder 37, JGA.

CHAPTER 11: THE EBB AND FLOW OF POLITICAL LIFE

1 Drew, *Campaign Journal*, 250.

2 John Glenn, remarks of Senator John Glenn before the Goddard Memorial Dinner, National Space Club, March 27, 1981, 5–6, Senate Papers Subgroup, Box 199, Folder 22, JGA.

3 John H. Glenn Jr., letter to Jake Garn, April 15, 1982, 1–2, Senate Correspondence (1981–84) Folder, Record 801, HRC.

4 Grimes, interview by author.

5 Michael Ryan, "Presidential Aspirations: In 1982, in the Early Stages of Glenn's Campaign for Higher Office People Magazine Took the Measure of a Promising Candidate," in Gibbs, "John Glenn," 56–59.

6 Howell Raines and Special to the *New York Times*, "Kennedy, Barring '84 Race, Cites Duty to Children," *New York Times*, December 2, 1982.

7 Jack W. Bermond and Jules Witcover, *Wake Us When It's Over: Presidential Politics in 1984* (New York: Macmillan, 1985), 42.

8 Glenn, oral history interview 23, Thomas, 26.

9 Steven Pearlman, letter to Susan Carnohan, March 18, 1983, 2, John Glenn Archives, Senate Papers Box 202, Folder 12, JGA.

10 Martin Schram, "Glenn and Mondale Collect a Set of Endorsements," *Washington Post*, June 9, 1983, Glenn, John H. Jr., Presidential Race, (1 of 3) Folder, HRC.

11 Glenn, oral history interview 23, Thomas, 16.

12 "Outline for Announcement Swing," undated, Senate Papers Subgroup, Box 89, Folder 6, JGA.

13 Howard Kurtz and Mary Thornton, "Glenn's Earnings Soar above Those of His Rivals for Nomination," *Washington Post*, May 18, 1983.

14 Martin Schram, "Glenn Even with Mondale for First Time in a Presidential Poll," *Washington Post*, May 15, 1983.

15 Howell Raines, "Glenn Seeking to Turn a Hero's Image into Votes," *New York Times*, June 15, 1983.

16 "Mondale and Glenn Lead in Survey of Democrats," *New York Times*, July 3, 1983.

17 Dr. Sawyer & Associates, Ltd., "John Glenn: Prospects, Problems and Positioning for the 1984 Primaries and General Election," Kennan Research & Consulting, June 1983, 27, Senate Papers Subgroup, Box 124, Folder 35, JGA.

18 John Glenn, "Report: On the Economy," Senate Papers Subgroup, Box 107, Folder 34, JGA.

19 John Glenn, "Report: On the Foreign Trade," Senate Papers Subgroup, Box 107, Folder 34, JGA.

20 John Glenn, "Report: On the Industrial Policy," Senate Papers Subgroup, Box 107, Folder 34, JGA.

21 John Glenn, "Report: On Agriculture," Senate Papers Subgroup, Box 107, Folder 34, JGA.

22 John Glenn, "Report: On Energy," Senate Papers, Box 107, Folder 34, JGA.

23 Balz, "Methodical Hero."

24 Butland, interview by author.

25 Peter Goldman and Tony Fuller, *The Quest for the Presidency 1984* (New York: Bantam Books, 1985), 77.

26 Spencer Rich, "Glenn to Focus Effort on Winning Delegates, Not States' Straw Polls, *Washington Post*, July 31, 1983.

27 Butland, interview by author.

28 Martin Schram, "Glenn's Campaign Staff Is Off and Walking," *Washington Post*, April 26, 1983.

29 Warren Wheat, "Glenn Turns Polls into Newest Launching Pad," *USA Today*, November 23, 1983, Senate Papers Subgroup, Box 124, Folder 51, JGA.

30 "Liftoff for Campaign 1984," *Newsweek*, October 3, 1983, 30.

31 James Reston, "The Wrong Stuff," *New York Times*, November 20, 1983.

32 Bermond and Witcover, *Wake Us*, 58.

33 Tom Sherwood, "Gov. Robb Endorses Glenn," *Washington Post*, December 6, 1983.

34 Bermond and Witcover, *Wake Us*, 104–105.

35 John Dillin, "Parrying Between Glenn and Mondale Heats Up Campaign," *Christian Science Monitor*, October 20, 1983, www.csmonitor.com/1983/1020/102018.html.

36 Bermond and Witcover, *Wake Us*, 106.

37 Glenn, oral history interview 23, Thomas, 22.

38 Susan Farrell, Susan Archer, and Josephine Sedgwick, "John Glenn, Space and Hollywood," Newsclip, *New York Times*, December 9, 2016, www.nytimes.com/video /multimedia/100000004813429/john-glenn-space-and-hollywood.html.

39 Butland, interview by author.

40 Glenn, oral history interview 23, Thomas, 19.

41 Glenn, oral history interview 23, Thomas, 19

42 Drew, *Campaign Journal*, 188–189.

43 Bermond and Witcover, *Wake Us*, 119.

44 Glenn, oral history interview 23, Thomas, 18.

45 Drew, *Campaign Journal*, 267.

46 Bill Kling, "Glenn's Drive to Centerfield Not Getting Him to First Base," *Washington Times*, January 23, 1984.

47 James J. Kilpatrick, "A Little Lead in the Saddlebags," *Seymour (IN) Daily Tribune*, November 2, 1983.

48 Martin Schram, "DEFLATION: Glenn's 'Spirit of Excellence' Gets Lost in Rough and Tumble," *Washington Post*, February 23, 1984.

49 Drew, *Campaign Journal*, 350.

50 Jonathan Moore, ed. *Campaign for President: The Managers Look at '84* (Dover, MA: Auburn House, 1986), 277, 278.

51 Moore, *Campaign for President*, 260.

52 Keith Blume, *The Presidential Election Show: Campaign '84 and Beyond on the Nightly News* (South Hadley, MA: Bergin & Garvey, 1985), 20.

53 Connolly, interview by author.

54 Glenn, oral history interview 23, Thomas, 17–18.

55 Glenn, oral history interview 23, Thomas, 15.

56 Glenn, oral history interview 23, Thomas, 18.

57 John Jacobs, "Wild Spending That Drained Glenn's Campaign Fund," *San Francisco Examiner*, May 13, 1984.

58 Butland, interview by author.

59 Federal Election Commission, "Interim Report of the Audit Division on the John Glenn Presidential Committee Inc.," January 17, 1985, 22, Senate Papers Subgroup, Box 86, Folder 14, JGA.

60 Glenn, oral history interview 23, Thomas, 32.

61 William Hershey, "Glenn Debt Still Grows," *Akron Beacon Journal*, July 23, 1991.

62 Glenn, oral history interview 23, Thomas, 32.

63 Annie Glenn, "Testimony of Annie Glenn Before the Platform Committee of the Democratic National Committee," May 21, 1984, 1, Senate Papers Subgroup, Box 51, Folder 1, JGA.

64 Butland, interview by author.

65 Moore, *Campaign for President*, 279–280.

66 Glenn, oral history interview 22, Thomas, 10.

67 "John Glenn: 'We Have a Tragedy That Goes Along with Our Triumphs,' Challenger Disaster—1986," Speakola, January 28, 1986, https://speakola.com/ideas/john-glenn -challenger- disaster-1986.

68 William G. Tull, letter to John Glenn, April 21, 1986, Senate Papers Subgroup, Box 48, Folder 9, JGA.

69 "The Space Shuttle Children's Trust Fund," Charity Navigator, 2017, www.charitynavigator.org/index.cfm?bay=search.profile&ein=521439509.

70 John Glenn, "Statement of Senator John Glenn on the Space Shuttle 'Challenger' Tragedy," January 28, 1986, 1, Senate Papers Subgroups, Box 199, Folder 2, JGA.

71 Grimes, interview by author.

72 Judith Havemann, "'Mr. Checklist's' Busy Agenda," *Washington Post*, November 26, 1986.

73 Grimes, interview by author.

74 *Report of the Committee on Governmental Affairs, U.S. Senate* (Washington, DC: Government Printing Office, 1989), 2.

75 Jon Margolis, "Heroic John Glenn Could Be Just Ticket for Dukakis' Psyche," *Chicago Tribune*, July 5, 1988.

76 Douglas Applegate, letter to John Glenn, June 22, 1988, Senate Papers Subgroup, Box 69, Folder 2, JGA.

77 Paul Light, memo to Senator Glenn, June 20, 1988, Senate Papers Subgroup, Box 69, Folder 2, JGA.

78 Donald Lambro, "Glenn Moves Up as Dukakis Trims List for Running Mate," *Washington Times*, July 4, 1988.

79 Abraham H. Miller, "The IRS Ten: The New Keating Five," *American Spectator*, June 13, 2013, https://spectator.org/55420_throw-book-him.

80 Glenn with Taylor, *John Glenn*, 469.

81 Helen Dewar, "Ethics Panel Hears 'Keating 5' Charges," *Washington Post*, November 16, 1990.

82 Grimes, interview by author.

83 John Glenn, "Glenn Statement on Presidential Address," November 30, 1990, Senate Papers Subgroup, Box 207, Folder 16, JGA.

84 "Confrontation in the Gulf: Roll Call in Senate on Gulf Assault," *New York Times*, January 14, 1991.

85 R. J. Reinhart, "George H. W. Bush Retrospective, *Gallup*, December 1, 2018, https://news.gallup.com/opinion/gallup/234971/george-bush-retrospective.aspx.

86 John Glenn, "Summary Memorandum—Trip to Saudi Arabia, Kuwait, and Iraq," March 14– 18, 1991, 1–2, Senate Papers Subgroup, Box 207, Folder 16, JGA.

87 John Glenn, "Opening Statement of Senator John Glenn, Subcommittee on Military Readiness and Defense Infrastructure," April 18, 1994, 2–3, Senate Papers Subgroup, Box 204, Folder 15, JGA.

88 Mike DeWine, "Lt. Governor Mike DeWine's Announcement Speech for the United States Senate," February 3, 1992, 3, Senate Papers Subgroup, Box 159, Folder 43, JGA.

89 John Glenn, "John Glenn's Clean Campaign Pledge," January 31, 1992, Senate Papers Subgroup, Box 159, Folder 43, JGA.

90 Ken Ringle, "John Glenn's Wavering Orbit," *Washington Post*, October 22, 1992.

91 John Glenn, "Glenn Has No Accomplishments (Or: 'What Are Your Major Accomplishments?')," Senate Papers Subgroup, Box 159, Folder 10, JGA.

92 Haseley, interview by author.

93 "Just Who Supported Whom," *Akron Beacon Journal*, November 5, 1992.

94 Government Accounting Office: Account and Financial Management Division, "The Chief Financial Officers Act: A Mandate for Federal Financial Management Reform," September 1991, 6, Senate Papers Subgroup, Box 63, Folder 51, JGA.

95 Jason Deparle, "Rant, Listen, Exploit, Learn, Scare, Help, Manipulate, Lead," *New York Times Magazine*, January 28, 1996, www.nytimes.com/1996/01/28/magazine/rant-listen-exploit-learn-scare-help-manipulate-lead.html.

96 "About the OCWR," Office of Congressional Workplace Rights," www.ocwr.gov/about-ocwr.

97 Grimes, interview by author.

98 Albert R. Hunt, "A Bipartisan Coverup to Block an Explosive Inquiry," *Wall Street Journal*, March 6, 1997.

99 Grimes, interview by author.

100 Candy Crowley, "Senator John Glenn Talks about House Campaign Finance Investigation," July 2, 1997, All Politics (website),www.cnn.com/ALLPOLITICS/1997/07/02/thompson/transcript/index.html.

101 Haseley, interview by author.

102 Grimes, interview by author.

103 William J. Clinton, "Remarks for Gridiron Dinner," March 21, 1998, Clinton Presidential Records, Speechwriting, Jeff Shesol series, ID number 19942, Gridiron 3/21/98 Folder, WJCL.

CHAPTER 12: WINNING HIS WINGS AGAIN

1   John Glenn, "JSC Tapes," March 7, 1998 through April 15, 1998, transcript, 1–3, Non-Senate Papers Subgroup, Box 77, Folder 17, JGA.

2   Charles, interview by author.

3   "John Glenn to Fly in Space Again," news release, NASA, January 17, 1998, Clinton Presidential Records, Speechwriting, Lowell Weiss, Space Station Folder, WJCL.

4   Glenn, "JSC Tapes," 5.

5   L. Warren E. Leary, "Glenn Kept After NASA, and After Two Years the Agency Said Yes," *New York Times*, January 17, 1998.

6   Glenn, JSC Tapes, 6–7.

7   John Glenn, "Ask an Astronaut," National Space Society, 1996, 6, Glenn, John H. Jr., (1997– 2002) Folder, Record 799, HRC.

8   Glenn with Taylor, *John Glenn*, 483–484.

9   Dan Goldin, "Announcement of NASA's Decision to Fly John Glenn," National Aeronautics and Space Administration, January 16, 1998, 1.

10 Dan Goldin, "Announcement of NASA's Decision to Fly John Glenn," National Aeronautics and Space Administration, January 16, 1998, 1, Clinton Presidential Records, Speechwriting, Lowell Weiss Series, ID Number 17193, Folder Title Space Station, WJCL.

11 Seth Borenstein and Tamara Lytle, "Glenn Still Shows the Right Stuff; Space Encore OK'd—Teacher Will Train," *Orlando Sentinel*, January 17, 1998.

12 Ralph Vartabedian, "Glenn Dedicates Orbital Encore to Older Americans," *Los Angeles Times*, January 17, 1998.

13 Leary, "Glenn Kept."

14 Shepard, oral history interview, Neal, 30.

15 Katharine Q. Seelye, "Glenn Seeks to Slip the Bonds of Age in Space," *New York Times*, January 17, 1998.

16 Leary, "Glenn Kept."

17 Glenn, "JSC Tapes," 9–20.

18 Bill Clinton, "State of the Union: Transcript of the State of the Union Message from President Clinton," January 28, 1998, A19, Clinton Presidential Records, Speechwriting, Michael Waldman Series, ID Number 14460, Folder Title State of the Union 1998-Drafts, Folder 1, WJCL.

19 Glenn with Taylor, *John Glenn*, 488–489.

20 Glenn, "JSC Tapes," 22–28.

21 Glenn, "JSC Tapes," 77.

22 Glenn, "JSC Tapes," 70, 89.

23 Charles, interview by author.

24 Robinson, interview by author.

25 Glenn, "JSC Tapes," 31–34.

26 Glenn, "JSC Tapes,"53–56.

27 Glenn, "JSC Tapes,"38–40.

28 Glenn, "JSC Tapes,"57–58.

29 Glenn, "JSC Tapes,"41–42.

30 Charles, interview by author.

31 On his Mercury flight, Glenn carried an Ansco Autoset camera that he bought with his own money at a Cocoa Beach drugstore. See Alice George, "Two Decades Ago, John Glenn's Encore Space Flight Lifted US Spirits," *Smithsonian Magazine*, October 17, 2018, www.smithsonianmag.com/smithsonian-institution/two-decades-ago-john-glenns-encore -space-flight-lifted-us-spirits-180970512.

32 Glenn, "JSC Tapes," 47–48.

33 Robinson, interview by author.

34 Glenn, "JSC Tapes," 65–68, 78–79.

35 "Living in the Space Shuttle," NASA, FS-1995–[20]08—001JSC, Non-Senate Papers Subgroup, Box 78, Folder 41, JGA.

36 Alice George, "Rita Rapp Fed America's Space Travelers," *Smithsonian Magazine*, March 27, 2019, www.smithsonianmag.com/smithsonian-institution/rita-rapp-fed-americas -space-travelers-180971801.

37 Greenwich Mean Time is a time zone used in some European and African nations, and it matches Coordinated Universal Time, which provides a worldwide standard time.

38 Glenn with Taylor, *John Glenn*, 508.

39 Associated Press, "Joe Biden Joins Thousands of Mourners at Ohio Funeral for John Glenn," *New York Daily News*, December 17, 2016, https://www.nydailynews.com/news /national/joe-biden-joins-thousands-mourners-john-glenn-funeral-article-1.2914571.

40 B. Drummond Ayres Jr., "Political Briefing: Glenn Is Nominated for Box of Wheaties," *New York Times*, June 7, 1998, Section 1, Page 20, www.nytimes.com/1998/06/07/us /political-briefing-glenn-is-nominated-for-box-of-wheaties.html.

41 Glenn, "JSC Tapes," 82–85.

42 "To: Senator and Astronaut John Glenn," National Air and Space Museum, Non-Senate Papers Subgroup, Box 84, Folder 1, JGA.

43 Glenn with Taylor, *John Glenn*, 509.

44 Jeffrey Kluger, "Rocket Man Redux," in Gibbs, "John Glenn," 82.

45 Schirra, oral history interview, Neal, 50.

46 DeLeon, interview by author.

47 Calvin L. Burch, "STS-95 Security Measures," Kennedy Space Center, undated, 8, NASA Collection, Daniel Goldin Papers, Box 180, Folder 77576, NA/II.

48 Gordon Cooper with Bruce Henderson, *Leap of Faith: An Astronaut's Journey into the Unknown* (New York: HarperTouch, 2000), 297.

49 Messages left at Astronaut Hall of Fame, October 29, 1998, Non-Senate Papers Subgroup, Box 83, Folder 15, JGA.

50 Carpenter, oral history interview 2, Neal, 7.

51 "One Man, Two Missions," *Washington Post*, October 25, 1998.

52 "Spacecraft Comparison: Mercury/Friendship 7 and Space Shuttle Discovery," NASA, accessed October 12, 2017, https://spaceflight.nasa.gov/shuttle/archives/sts-95/veh _comparison.html.

53 "John Glenn's Shuttle Flight: STS 95 Highlights," YouTube video, posted by "Waspie_ Dwarf," December 10, 2016, https://youtu.be/A5arx7CAhew?t=1100.

54 Newcott, "John Glenn," 75.

55 "Excerpts from John Glenn's Greeting During Perth, Australia Flyover," October 30, 1998, Non-Senate Papers Subgroup, Box 79, Folder 2, JGA.

56 Newcott, "John Glenn," 61.

57 Paul E. Wagner, "Glenn's Dream Lives in our National Soul," *Houston Chronicle*, November 1, 1998.

58 Tom Shales, "A Booster Rocket Sends a Scandal-Weary Nation's Spirits Soaring," *Washington Post*, October 30, 1998.

59 "Transcript of Senator Glenn's First Remarks from Space," Kennedy Space Center Public Affairs, October 29, 1998, Non-Senate Papers Subgroup, Box 79, Folder 2, JGA.

60 Scott Parazynski with Suzy Flory, *The Sky Below: A True Story of Summits, Space, and Speed* (New York: Little A, 2017), 105.
61 "STS-95, Mission Control Center, Status Report #4," NASA, October 30, 1998, https://spaceflight.nasa.gov/spacenews/reports/sts95/STS-95-04.html.
62 "STS-95, Mission Control Center, Status Report #8," NASA, November 1, 1998, https://spaceflight.nasa.gov/spacenews/reports/sts95/STS-95-08.html.
63 Ronald L. Phelps, "Operational Experiences with the Petite Amateur Navy Satellite— PANSAT," Fifteenth Annual / Utah State University Conference on Small Satellites, 2001, https://digitalcommons.usu.edu/smallsat/2001/All2001/33.
64 "STS-95 Human Life Sciences Overview, Presented to Dr. Arnauld Nicogossian," April 24, 1998, Non-Senate Papers Subgroup, Box 75, Folder 42, JGA.
65 "STS-95, Mission Control Center, Status Report #9," NASA, November 2, 1998, https://spaceflight.nasa.gov/spacenews/reports/sts95/STS-95-09.html.
66 "STS-95, Mission Control Center, Status Report #17," NASA, November 6, 1998, https://spaceflight.nasa.gov/spacenews/reports/sts95/STS-95-17.html.
67 "Inflight Event: Interviews with U.S. Media," NASA, November 2, 1998, https://spaceflight.nasa.gov/shuttle/archives/sts-95/usconf.html.
68 "Inflight Event: STS-95 Educational Downlink," NASA, October 31, 1998, https://spaceflight.nasa.gov/shuttle/archives/sts-95/eduevent.html.
69 Newcott, "John Glenn," 80.
70 "STS-95, Mission Control Center, Status Report #5," NASA, October 31, 1998, https://spaceflight.nasa.gov/spacenews/reports/sts95/STS-95-05.html.
71 "STS-95, Mission Control Center, Status Report #7," NASA, November 1, 1998, https://spaceflight.nasa.gov/spacenews/reports/sts95/STS-95-07.html.
72 "Thought Leaders Series: John Glenn's Second Flight," Space Station Houston Facebook Live, November 8, 2018, www.facebook.com/SpaceCenterHouston/videos/116209499304428.
73 "STS-95, Mission Control Center, Status Report #02," NASA, October 29, 1998, https://spaceflight.nasa.gov/spacenews/reports/sts95/STS-95-02.html.
74 "STS-95, Mission Control Center, Status Report #9," NASA, November 2, 1998, https://spaceflight.nasa.gov/spacenews/reports/sts95/STS-95-09.html.
75 "STS-95, Mission Control Center, Status Report #11," NASA, November 3, 1998, https://spaceflight.nasa.gov/spacenews/reports/sts95/STS-95-11.html.
76 "Status Report #11."
77 Scott Parazynski and John Glenn, email to Annie Glenn, "FAMPS207," Non-Senate Papers Subgroup, Box 76, Folder1 7, JGA.
78 John Glenn, email to Annie Glenn, "Famps204," Non-Senate Papers Subgroup, Box 76, Folder 17, JGA.
79 John Glenn, "Message to President Bill Clinton from Senator John Glenn," November 6, 1998, WJC-ARMS Collection, Presidential Electronic Mail from the Automated Records Management System, 1/20/1993–1/20/2001, Automated Records Management System Email from the White House Office Bucket 1/20/1993–1/20/2001, WJCL.
80 Butland, interview by author.
81 "STS-95, Mission Control Center, Status Report #20," NASA, November 7, 1998, https://spaceflight.nasa.gov/spacenews/reports/sts95/STS-95-20.html.
82 Tribune News Services, "Glenn Comes Clean on Shuttle Landing," *Chicago Tribune*, December 2, 1998, www.chicagotribune.com/news/ct-xpm-1998-12-02-9812020083-story.html.
83 T. J. Milling, "Hero's Welcome: Houstonians Greet Glenn, Discovery Crew," *Houston Chronicle*, November 9, 1998.
84 Douglas Martin, "In Line of Parades, One for Glenn and the Crew of Discovery," *New York Times*, November 17, 1998.
85 Lisa Meyer, "Glenn Puts Half-Million New Yorkers into Orbit," *Los Angeles Times*, November 17, 1998.

86 Chip Johnson, "Berkley's Glenn Clan Exhales," *SFGate*, November 17, 1998, www.sfgate
    .com/bayarea/johnson/article/Berkeley-s-Glenn-Clan-Exhales-Son-just-glad-2978309
    .php.

87 "STS-95 Crew Appearance Requests," NASA, November 9, 1998, Non-Senate Papers
    Subgroup, Box 76, Folder 45, JGA.

88 Pauline Arrillaga, "Thousands Come Out to Honor the World's Oldest Astronaut," Asso-
    ciated Press, November 12, 1998.

89 Newcott, "John Glenn," 81.

90 John Charles, memo to Senator John Glenn, August 20, 1999, Non-Senate Papers Sub-
    group, Box 76.1, Folder 1, JGA.

91 Adrian LeBlanc et al., "Senator Glenn's Results for STS-95 Experiment Entitled 'Mag-
    netic Resonance Imaging (MRI) after Exposure to Microgravity,'" undated, Non-Senate
    Papers Subgroup, Box 76.1, Folder 8, JGA.

92 Paloski et al., "Abstract," undated, Non-Senate Papers Subgroup, Box 76.1, Folder 10,
    JGA.

93 Alfred C. Rossum, Margie L. Wood, and Janice Fritch-Yelle, "Evaluation of STS-95
    Cardiovascular Data," undated, Non-Senate Papers Subgroup, Box 76.1, Folder 5, JGA.

94 Raymond Stowe, letter to John Glenn, August 25, 1999, Non-Senate Papers Subgroup,
    Box 76.1, Folder 6, JGA.

95 Charles A. Czeisler, "Sleep, Respiration, and Melatonin in Microgravity," undated,
    Non-Senate

96 "Inflight Event: STS-95 EducationalDownlink."

97 Charles, interview by author.

98 Roger Rosenblatt, "A Realm Where Age Doesn't Count," *Time*, August 17, 1998, Glenn,
    John H. Jr., Return to NASA (8 of 9) Folder, Record 7885, HRC.

EPILOGUE

1 Newcott, "John Glenn," 81.

2 Robinson, interview by author.

3 John Glenn, "Remarks of Senator John Glenn," February 20, 1997, 13, Senate Papers
   Subgroup, Box 48, Folder 6, JGA.

4 Glenn, "Remarks," 8.

5 "Glenn Honored with New Stamp," NASA, *Dryden X-Press*, April 14, 2000, 1, Glenn,
   John H. Jr., Return to NASA (9 of 9) Folder, Record 16401, HRC.

6 Glenn, oral history interview 23, Thomas, 38.

7 Tim Russert, interview with John Glenn, *Meet the Press*, February 2, 2003, 13–14,
   Non-Senate Papers Subgroup, Box 57, Box 21, JGA.

8 Margaret J. Fargo, letter to John Glenn, undated, Non-Senate Papers Subgroup, Box
   105.2, Folder 28, JGA.

9 John Glenn, "Testimony Before the House of Representatives Committee on Science and
   Technology," July 30, 2008, 3–7, Non-Senate Papers Subgroup, Box 104, Folder 34, JGA.

10 "Glenn Sees U.S. Moving Backward in Space," *Aviation Week*, June 23, 2010, Glenn, John
    H. Jr. (2003–) Folder, Record 81101, HRC.

11 John Glenn, "A Century Later America Still Soars," *Parade Magazine*, June 29, 2003, 4.

12 Grimes, interview by author.

13 Domestic Policy Council and Caroline Chang, "News / Background on Administration's
    Education Priorities: Glenn Commission Report," September 27, 2000, 3, Clinton Digital
    Library, https://clinton.presidentiallibraries.us/items/show/51994.

14 John Glenn, foreword to *Report to the Nation from the National Commission on Mathematics
    and Science Teaching for the 21st Century*, 2000, Non-Senate Papers Subgroup, Box 104.1,
    Folder 27, JGA.

15 "H.R. 1585, Sec. 1062, Congressional Commission on the Strategic Posture of the United
    States," John Glenn, Non-Senate Papers Subgroup, Box 102.1, Folder 12, JGA.

16 Glenn, oral history interview 22, Thomas, 18.

17 Baker Spring, "Strategic Posture Commission's Report Provides Necessary Guidance to Congress," Heritage Foundation, May 13, 2009, www.heritage.org/defense/report /strategic-posture-commissions-report-provides-necessary-guidance-congress.

18 Associated Press, "John Glenn Enjoyed 'Frasier' Role," March 7, 2001, Glenn, John H. Jr. (1997–2002) Folder, Record 799, HRC.

19 Script, *Frasier*, December 21, 2000, 10, Non-Senate Papers Subgroup, Box 102.2, Folder 36, JGA.

20 Autographed script cover, *Frasier*, Paramount, Non-Senate Papers Subgroup, Box 102.2, Folder 36, JGA.

21 Peter Nero, script for John Glenn, "Voyage into Space," Dayton Philharmonic Orchestra, November 18, 2005, Non-Senate Papers Subgroup, Box 110, Folder 7, JGA.

22 David Fahey, interview of John Glenn, October 5, 2009, 6, Non-Senate Papers Subgroup, Box 110, Folder 98, JGA.

23 Sherrod Brown, "Commending John Hershel Glenn," *Congressional Record—Senate*, July 18, 2011, S4635.

24 Harry Reid, "John Glenn," *Congressional Record—Senate*, July 18, 2011, S4635.

25 Neil Armstrong, email to John Glenn, November 5, 2011, Non-Senate Papers Subgroup, Box 102.2, Folder 2, JGA.

26 "John Glenn Given Medal of Freedom," National Aeronautics and Space Administration, October 31, 2013, www.nasa.gov/multimedia/imagegallery/image_feature_2264.html.

27 John Glenn, "Scott Carpenter Memorial Service," November 2, 2013, 3, Non-Senate Papers Subgroup, Box 11.5, Folder 16, JGA.

28 Glenn, "Carpenter Memorial Service," 19.

29 Astronaut Scholarship Foundation, accessed February 24, 2019, https:// astronautscholarship.org/index.html.

30 John Glenn, Calendar 2014, Non-Senate Papers Subgroup, Box 115, Folder 19, JGA.

31 Jay Barbree, "FRM. Astronaut John Glenn Dead at Age 95," NBC News, December 8, 2016, www.nbcnews.com/video/nbc-s-jay-barbree-reflects-on-friendship-with-john -glenn-828322371602.

32 Marla Matzer Rose, "John Glenn Honored as Columbus Airport Is Renamed for Him," *Columbus Dispatch*, June 28, 2016, www.dispatch.com/content/stories/business /2016/06/28/0628-john-glenn-honored-at-airport-renaming-ceremony.html.

33 Mattis served as secretary of defense from the beginning of the Trump administration in January 2017 until December 2018, when he resigned after Trump countermanded his commitment to keep US troops in Syria.

34 DeLeon, interview by author.

35 Sherrod Brown, "Brown Statement on Passing of Senator John Glenn," Ohio senator Sherrod Brown's website, December 8, 2016, www.brown.senate.gov/newsroom/press /release/brown-statement-on-passing-senator-john-glenn.

36 Rob Portman, "The John Glenn I Knew," Ohio senator Rob Portman's website, December 12, 2016, www.portman.senate.gov/newsroom/columns/john-glenn-i-knew.

37 Mitch McConnell, "McConnell Remembers John Glenn: 'Today Our Nation Bids Farewell to One of the Most Iconic Figures of the 20th Century,'" press release posted on Senate Majority Leader Mitch McConnell's website, December 8, 2016, www .republicanleader.senate.gov/newsroom/press-releases/mcconnell-remembers-john-glenn.

38 Margot Lee Shetterly, *Hidden Figures: The American Dream and the Untold Story of the Black Women Mathematicians Who Helped Win the Space Race* (New York: William Morrow, 2016), 214–217, 223–225.

39 Jim Otte, "Thousands Pay Tribute to Glenn," *Springfield (OH) News*, December 17, 2016.

40 William Harwood, "John Glenn Recalled as Humble Hero, and 'Just Dad' to His Children," CBS News, December 17, 2016, www.cbsnews.com/news/john-glenn-remembered -as-humble-hero-and-just-dad.

41 Harwood, "Recalled as Humble Hero."

42 Harwood.

43 Mark McCarter, "John Glenn Was a National Treasure," *Anniston (AL) Star*, December 22, 2016.

44 Barbree, "John Glenn Dead."

45 Harwood, "Recalled as Humble Hero."

46 Michael E. Ruane, "A Hero's Rainy Farewell: John Glenn Is Buried at Arlington with Marine Corps Ceremony," *Philadelphia Inquirer*, April 7, 2017.

47 "Godspeed, John Glenn!" *New York Times*, April 6, 2017, www.nytimes.com/2017/04/06/us/john-glenn-arlington-national-cemetery.html.

48 Ruane, "Hero's Rainy Farewell."

49 Robinson, interview by author.

50 Parazynski with Flory, *Sky Below*, 111.

51 Robinson, interview by author.

52 Brenna Busby, "Glenn Historic Site Becomes Museum," OrbitMediaOnline.com, August 27, 2016, www.orbitmediaonline.com/glenn-museum-changes-hands.

53 "Glenn's House Going Downtown," *Times Recorder*, February 23, 2001, 1.

54 "Glenn Research Center," National Aeronautics and Space Administration, accessed February 24, 2019, www.nasa.gov/centers/glenn/home/index.html.

55 John Glenn Astronomy Park, accessed February 24, 2019, https://jgap.info.

56 Marcia Dunn, "Glenn Supply Ship at Space Station," Philadelphia Inquirer, April 23, 2017.

57 Elizabeth Howell, "New Glenn: Blue Origin's Reusable Rocket," Space.com, May 1, 2018, www.space.com/40455-new-glenn-rocket.html.

58 Robert Pearlman, "One-of-a-Kind Exhibit on NASA's Mercury Space Program Opens at Grand Turk Cruise Center," collectSPACE, January 31, 2011, www.collectspace.com/ubb/Forum41/HTML/000390.html.

59 McDougall, *Heavens and the Earth*, 347.

60 Phil McCausland and Jon Schuppe, "'The Last National Hero': John Glenn Dead at 95," NBC News, December 9, 2016, www.nbcnews.com/news/us-news/last-true-national-hero-john-glenn-dead-95-n693391.

61 Wolfe, *Right Stuff*, 348.

62 DeLeon, interview by author.

63 Charles, interview by author.

64 DeLeon, interview by author.

65 Butland, interview by author.

66 Marshall Fishwick, "Post-Modern Blues," *Source Studies in Popular Culture*, April 2000, 9.

67 Robinson, interview by author.

BIBLIOGRAPHY

Bermond, Jack W., and Jules Witcover. *Wake Us When It's Over: Presidential Politics in 1984*. New York: Macmillan, 1985.

Blume, Keith. *The Presidential Election Show: Campaign '84 and Beyond on the Nightly News*. South Hadley, MA: Bergin & Garvey, 1985.

Boomhower, Ray E. *Gus Grissom: The Lost Astronaut*. Indianapolis: Indiana Historical Society Press, 2004.

Burrows, William E. *This New Ocean: The Story of the First Space Age*. New York: Modern Library, 1998.

Byrnes, Mark E. *Politics and Space: Image Making by NASA*. Westport, CT: Praeger, 1994.

Carpenter, Scott, with Kris Stoever. *For Spacious Skies: The Uncommon Journey of a Mercury Astronaut*. New York: New American Library, 2004.

Committee on Science and Astronautics, US House of Representatives, Eighty-Sixth Congress, Second Session. *The Practical Values of Space Exploration*. Washington, DC: Government Printing Office, 1960.

Cooper, Gordon, with Bruce Henderson. *Leap of Faith: An Astronaut's Journey into the Unknown*. New York: HarperTouch, 2000.

Cunningham, Walter. *The All-American Boys: An Insider's Look at the U.S. Space Program*. New York: ipicturebooks, 2009.

Dick, Stephen J., ed. *Remembering the Space Age: Proceedings of the 50th Anniversary Conference*. Washington, DC: National Aeronautics and Space Administration, 2008.

Dille, John. Introduction to *We Seven*, by M. Scott Carpenter, L. Gordon Cooper Jr., John H. Glenn Jr., Virgil L. Grissom, Walter M. Schirra Jr., Alan B. Shepard Jr., and Donald K. Slayton, 5–22. New York: Simon & Schuster, 1962.

Drew, Elizabeth. *Campaign Journal: The Political Events of 1983–1984*. New York: Macmillan, 1985.

Gibbon, Peter H. *A Call to Heroism: Renewing America's Vision of Greatness*. New York: Grove, 2007.

Gibbs, Nancy, ed. "John Glenn: A Hero's Life, 1921–2016." Commemorative edition, *Time*, 2016.

Glenn, John. *P.S. I Listened to Your Heart Beat: Letters to John Glenn*. Houston: World Book Encyclopedia Science Service, 1964.

Glenn, John, with Nick Taylor. *John Glenn: A Memoir*. New York: Bantam Books, 1999.

Goldman, Peter, and Tony Fuller. *The Quest for the Presidency 1984*. New York: Bantam Books, 1985.

Grimwood, James M. *Project Mercury: A Chronology*. Washington, DC: National Aeronautics and Space Administration, 1963.

Grosser, George H., Henry Wechsler, and Milton Greenblatt, eds. *The Threat of Impending Disaster*. Cambridge: Massachusetts Institute of Technology, 1964.

Hampton, Dan. *The Flight: Charles Lindbergh's Daring and Immortal 1927 Transatlantic Flight*. New York: HarperCollins, 2017.

Hersch, Matthew H. *Inventing the American Astronaut*. New York: Palgrave Macmillan, 2012.

Hilty, James W. *Robert Kennedy: Brother Protector*. Philadelphia: Temple University Press, 1997.

Kauffman, James L. *Selling Outer Space: Kennedy, the Media, and Funding for Project Apollo, 1961–1963*. Tuscaloosa: University of Alabama Press, 1994.

Kluger, Jeffrey. *Apollo 8: The Thrilling Story of the First Mission to the Moon*. New York: Henry Holt, 2017.

Koppel, Lily. *The Astronaut Wives Club: A True Story*. New York: Grand Central, 2013.

Kraft, Chris. *Flight: My Life in Mission Control*. New York: Penguin Books, 2001.

Lovell, Jim, and Jeffrey Kluger. *Apollo 13*. Boston: Houghton Mifflin, 1994.

McDougall, Walter A. *The Heavens and the Earth: A Political History of the Space Age*. Baltimore: Johns Hopkins University Press, 1997.

Mersky, Peter B. *U.S. Marine Corps Aviation Since 1912*. Annapolis: Naval Institute Press, 2009.

Montville, Leigh. *Ted Williams: The Biography of an American Hero*. New York: Broadway Books, 2005.

Moore, Jonathan, ed. *Campaign for President: The Managers Look at '84*. Dover, MA: Auburn House, 1986.

Parazynski, Scott, with Suzy Flory. *The Sky Below: A True Story of Summits, Space, and Speed*. New York: Little A, 2017.

Reeves, Richard. *President Kennedy: Profile of Power*. New York: Simon & Schuster, 1993.

Report of the Committee on Governmental Affairs, U.S. Senate. Washington, DC: Government Printing Office, 1989.

Schirra, Wally, with Richard N. Billings. *Schirra's Space*. Annapolis: Naval Institute Press, 1988.

Schlesinger, Arthur M., Jr. *Robert Kennedy and His Times*. Boston: Houghton Mifflin, 1978.

Seidel, Michael. *Ted Williams: A Baseball Life*. South Orange, NJ: Summer Games Books, 2015.

Shepard, Alan, and Deke Slayton with Jay Barbee. *Moon Shot*. New York: Open Road, 2011.

Shetterly, Margot Lee. *Hidden Figures: The American Dream and the Untold Story of the Black Women Mathematicians Who Helped Win the Space Race*. New York: William Morrow, 2016.

Sorensen, Theodore C. *Kennedy*. New York: Harper & Row, 1965.

Swenson, Loyd S., Jr., James M. Grimwood, and Charles C. Alexander. *This New Ocean: A History of Project Mercury*. Washington, DC: National Aeronautics and Space Administration, 1998.

Thompson, Neal. *Light This Candle: The Life and Times of Alan Shepard*. New York: Three Rivers, 2005.

Weitekamp, Margaret A. *Right Stuff, Wrong Sex: America's First Women in Space Program*. Baltimore: Johns Hopkins University Press, 2004.

Wolfe, Tom. *The Right Stuff*. New York: Bantam Books, 1979.

INDEX

Note: Page numbers in italics refer
to images